Molecular Simulations of Energy Materials

Molecular Simulations of Energy Materials

Guest Editors

Viorel Chihaia
Godehard Sutmann

MDPI

Basel • Beijing • Wuhan • Barcelona • Belgrade • Novi Sad • Cluj • Manchester

Guest Editors
Viorel Chihaia
Institute of Physical Chemistry "Ilie Murgulescu"
Romanian Academy
Bucharest
Romania

Godehard Sutmann
Jülich Supercomputing Centre (JSC)
Forschungszentrum Jülich
Jülich
Germany

Editorial Office
MDPI AG
Grosspeteranlage 5
4052 Basel, Switzerland

This is a reprint of the Special Issue, published open access by the journal *Molecules* (ISSN 1420-3049), freely accessible at: https://www.mdpi.com/journal/molecules/special_issues/Molecular_Simulations_Energy_Materials.

For citation purposes, cite each article independently as indicated on the article page online and as indicated below:

Lastname, A.A.; Lastname, B.B. Article Title. *Journal Name* **Year**, *Volume Number*, Page Range.

ISBN 978-3-7258-5909-2 (Hbk)
ISBN 978-3-7258-5910-8 (PDF)
https://doi.org/10.3390/books978-3-7258-5910-8

Contents

About the Editors

Viorel Chihaia

Viorel Chihaia graduated from the Faculty of Physics, University of Bucharest in 1989 and earned his Ph.D. in Chemistry in 1999 from the Ilie Murgulescu Institute of Physical Chemistry of the Romanian Academy (IPC), Bucharest. He has been affiliated with IPC since 1990, where he currently serves as a Senior Researcher.

Throughout his career, he has been engaged in research work at several international institutions, including Keimyung University (Daegu, South Korea, 2000), the University of Göttingen (Germany, 2001–2004), the Institute of Atomic and Molecular Sciences, Academia Sinica (Taiwan, 2010), and the Jülich Supercomputing Centre, Forschungszentrum Jülich (Germany, 2011–2015). He has coordinated and participated in numerous national and international scientific projects, notably serving as director of the national project "Virtual Group for Atomic Scale Simulations in Materials Science." He is currently involved in the High-Performance Computing (HPC) in the Physical Chemistry research program at IPC.

His research encompasses surface science, catalysis, and computational materials science, with particular expertise in quantum chemistry, solid-state modeling, molecular dynamics, and Monte Carlo simulations. He has developed HPC-based parallel algorithms and simulation methods for investigating phenomena across nano-, meso-, and multi-scale levels, whereas his current work focuses on the development and application of HPC simulation techniques for energy materials.

Godehard Sutmann

Godehard Sutmann has studied theoretical physics and has been working as a computational scientist in the field of high-performance parallel computing, applied mathematics, and method development. He has studied in Göttingen, Montpellier and Heidelberg and received his PhD in 1999 at the Technical University Munich. He has worked as a postdoctoral researcher at the University of Trento and continued his work at Forschungszentrum Jülich, Germany. NIH Washington, UC Santa Barbara and UNAM at Mexico City are, among others, institutions where he was appointed as a research scientist. His research interest is in computational methods and algorithms for complex many-particle physics and statistical physics with a main focus on parallel computing. At present, he works at the Jülich Supercomputing Centre (JSC) at Forschungszentrum Jülich, where he is the leader of the Simulation and Data Laboratory Complex Particle Systems. Additionally, he is a director of the German CECAM node at Forschungszentrum Jülich, where he organizes and coordinates workshops and tutorials within the CECAM calendar. Meanwhile, his second affiliation is at the Ruhr-University Bochum, as a member of the Interdisciplinary Centre for Advanced Materials Simulation (ICAMS), and where he also works as professor for High Performance Computing.

Preface

The transition toward sustainable energy systems represents one of the most pressing scientific and technological imperatives of the twenty-first century. The accelerating global demand for energy, together with the depletion of conventional fossil resources and the escalating environmental consequences of their use, underscore the necessity of developing innovative, sustainable alternatives. Addressing this multifaceted challenge requires a paradigm shift in material design—one that simultaneously advances energy generation, storage, and conversion while adhering to principles of environmental stewardship.

The inherent complexity of energy systems, which span broad spatial and temporal scales, necessitates the adoption of advanced methodologies capable of elucidating their underlying physicochemical mechanisms. In this context, molecular simulations have emerged as indispensable tools in the study and design of energy materials. By enabling the detailed examination of interactions at the atomic and molecular level, these computational approaches provide insights that complement and extend beyond those attainable through experimental techniques.

From elucidating electronic structures and reaction dynamics to modeling large-scale material behavior, molecular simulations furnish a versatile framework for predicting performance, optimizing design, and accelerating the discovery of novel energy materials. Their integration into the broader field of materials science has transformed our capacity to rationalize complex phenomena and to guide the development of efficient, high-performance, and sustainable energy solutions.

This Reprint brings together a collection of articles that highlight the transformative impact of computational methodologies on the advancement of energy materials science and engineering. Viewed through this perspective, the included works demonstrate how molecular-level insights derived from simulations can guide and accelerate the development of next-generation sustainable energy technologies.

Viorel Chihaia and Godehard Sutmann
Guest Editors

molecules

Editorial

Molecular Simulations of Energy Materials

Viorel Chihaia [1,*] and Godehard Sutmann [2,3,*]

1 Institute of Physical Chemistry "Ilie Murgulescu" of the Romanian Academy, Splaiul Independentei 202, Sector 6, 060021 Bucharest, Romania
2 Jülich Supercomputing Centre (JSC), Institute for Advanced Simulation (IAS), Forschungszentrum Jülich, 52428 Jülich, Germany
3 Interdisciplinary Centre for Advanced Materials Simulations (ICAMS), Ruhr Universität Bochum, 44801 Bochum, Germany
* Correspondence: vchihaia@icf.ro (V.C.); godehard.sutmann@rub.de (G.S.)

The accelerating demand for energy, coupled with the ongoing depletion of conventional energy resources and environmental problems, poses a critical challenge to the scientific community [1]. Addressing this challenge requires the development of innovative materials capable of generating, converting, storing, and utilizing energy in ways that are both sustainable and environmentally benign [2]. Understanding these complex systems—spanning diverse phenomena and interacting across multiple spatial (from atomic to macroscopic) and temporal (from femtoseconds to years) scales—demands an integrated scientific approach [3]. While experimental research remains essential in uncovering the behavior of energy materials, especially under harsh environmental conditions, many microscopic-scale mechanisms remain poorly understood [4]. This is where molecular-level computational simulations can play an important role. Advances in computer molecular sciences now offer powerful methods for probing the structure, dynamics, and reactivity of materials at the atomic and molecular levels, complementing experimental findings and offering predictive insights [5]. In particular, molecular simulations—encompassing static modeling, molecular dynamics, and Monte Carlo methods—enable the exploration of energy materials under various conditions [6]. These approaches can operate across quantum, classical, and coarse-grained frameworks, each providing valuable perspectives on intra- and intermolecular forces. Quantum mechanical methods reveal critical details of electronic structure, which underpin macroscopic properties and device performance [7], while atomistic and coarse-grained simulations offer scalable insights into larger systems and longer-time-scale processes [8]. To fully capture the multiscale nature of energy materials, there is a growing need to integrate particle-based methods with continuum models through multiresolution and multiscale approaches [9,10]. Such hybrid strategies promise to deepen our understanding of the fundamental phenomena governing the behavior of materials in real-world energy and environmental applications.

This Special Issue aims to highlight recent advances in atomic-scale simulation methods and their application to energy materials science. Contributions demonstrate how computational tools provide crucial insights into the design, characterization, and optimization of materials for a sustainable energy future. The main investigated properties and phenomena are summarized in Table 1.

Hydrogen production through solar-driven photoelectrochemical (PEC) water splitting represents one of the most promising pathways toward sustainable energy. However, traditional semiconductors such as titanium dioxide (TiO_2) face key limitations, such as a wide bandgap and poor conductivity. In this context, Akbar et al. (contribution 1) employ density functional theory (DFT) to explore how doping TiO_2 with transition metals

 https://doi.org/10.3390/molecules30214270

(Ag, Fe, and Co) can improve its visible-light absorption and charge transport. Their comparative analysis of dopants, using GGA+U and hybrid functionals, illustrates how atomistic modifications influence electronic structure, optical response, and mechanical stability—ultimately guiding the rational design of efficient photocatalysts.

Table 1. The properties, phenomena, and simulation methods involved in the articles of the present Special Issue dedicated to molecular simulations for energy materials.

Contribution	Topic	Properties Studied	Phenomena	Simulation Methods [a]
1	Doped TiO_2 (Ag, Fe, Co) for water splitting	Band gap, elastic constants, mechanical stability	Doping, water dissociation, light absorption, photoconductivity	DFT calculations
2	FeN_4-doped graphene nanoribbons	Spin distribution, electronic structure, magnetism	Effect of dopant position and edge termination on electronic/magnetic properties	Spin-polarized DFT
3	Mixed-metal oxide perovskites (BaCe–Fe–O)	Thermodynamic stability, surface reduction energy, oxygen vacancy behavior	Surface dehydration, H_2 interaction, defect formation	DFT-based thermodynamics
4	Xanthione in polar solvents	Electronic transitions, solvent sensitivity, photostability	Solvent effect on excited states, non-Kasha behavior	TDDFT, QC-MD, explicit and implicit solvent
5	Quantum battery network	Exciton population, energy retention, site energy influence	Dark state storage, discharge dynamics, exciton transfer	Open quantum network theory, quantum dynamics simulations
6	SAGD process–bitumen/quartz interaction	Wettability, adsorption energy, surface affinity	Wetting alteration under high temperature, asphaltene adsorption	CFF-MD, with varying conditions
7	HPAM polymer on calcite	Adsorption behavior, interaction strength, effect of ionic environment	Polymer adsorption, salt-bridging, charge screening	CFF-MD simulations
8	MOFs and zeolites for He/N_2 separation	Henry's constant, diffusion coefficients, selectivity, permeability	Gas adsorption, diffusion, and membrane-based separation	GCMC, Equilibrium CFF-MD

[a] see the short notations listed in the Abbreviations section.

Alongside efforts in hydrogen production, the search for cost-effective electrocatalysts for fuel cells remains a critical area of research. Fe–N–C systems, especially FeN_4-doped graphene nanoribbons (GNRs), have emerged as high-potential non-precious metal catalysts for oxygen reduction reactions (ORRs). Mineva et al. (contribution 2) present a DFT-based study on how dopant position, edge termination, and spin ordering affect the electronic and magnetic properties of FeN_4-GNR systems. Their work provides insights into spin polarization control and its implications for catalytic activity, highlighting potential applications in spintronic devices and magneto-electronic catalysis.

Hydrogen separation technologies are equally crucial in advancing the hydrogen economy. Rosen and Sohlberg (contribution 3) investigate dual-phase perovskite materials ($BaCe_{0.85}Fe_{0.15}O_{3-\delta}/BaCe_{0.15}Fe_{0.85}O_{3-\delta}$) using DFT and first-principles thermodynamics. Their work focuses on surface reduction, vacancy formation, and thermodynamic stability under hydrogen-rich conditions. By constructing Gibbs free energy profiles and analyzing defect behavior, their study informs the design of stable, high-performance materials for applications such as proton-conducting membranes and hydrogen purification systems.

The importance of solvation effects in photophysics is explored by Manian et al. (contribution 4), who focus on xanthione—a sulfur-containing polycyclic aromatic compound known for its unusual anti-Kasha behavior and high solvent sensitivity. Through a hybrid approach combining quantum chemistry and MD simulations, the authors assess how different solvation models (implicit vs. explicit) impact the molecule's excited-state dynamics. Their findings emphasize the necessity of explicit solvent modeling for accurately capturing the photophysical behavior of solvent-sensitive materials, with implications for quantum photonics, molecular electronics, and light-harvesting systems.

The intersection of quantum mechanics and energy storage is exemplified by Liu and Hanna (contribution 5), who explore quantum batteries—devices that utilize quantum coherence to store and transfer energy via excitonic mechanisms. Their simulation-based study introduces a symmetry-enabled open quantum network model, revealing how exchange symmetry and decoherence affect energy retention and transfer dynamics. This work opens new directions in quantum energy material research, bridging condensed matter physics and nanoscale energy design.

On the molecular scale, wettability and interfacial dynamics play a critical role in enhanced oil recovery (EOR). Ahmadi and Chen (contribution 6) use MD simulations to investigate how asphaltene concentration in oil droplets affects their adsorption on quartz surfaces, particularly under high-temperature conditions relevant to Steam-Assisted Gravity Drainage (SAGD). Their study uncovers the interplay between oil composition, temperature, and surface interactions, offering mechanistic insights into thermally induced wettability alteration and interfacial behavior in heavy oil reservoirs.

Subsurface challenges such as solid production in carbonate reservoirs can be addressed through atomistic modeling. Hue et al. (contribution 7) simulate the adsorption of hydrolyzed polyacrylamide (HPAM) on calcite surfaces using classical MD in isothermal–isobaric ensembles. Their results provide critical data for designing polymer-based additives to stabilize formation rock and prevent damage, optimizing EOR strategies under varying geochemical conditions.

Grenev and colleagues (contribution 8) conducted a computational screening study to identify metal–organic frameworks (MOFs) and zeolites with high potential for helium/nitrogen (He/N_2) separation—an increasingly important challenge in industrial gas purification. Given helium's vital role in fields such as cryogenics and nuclear technology, along with its growing scarcity, there is an urgent need for efficient, selective, and cost-effective separation materials. This study highlights the power of high-throughput molecular simulations in accelerating the discovery of porous materials with optimized pore architectures and sorption properties tailored to specific gas separation tasks. By screening over 10,000 MOFs and 218 zeolites, Grenev and colleagues demonstrated the ability to identify materials with superior adsorption and membrane selectivity for helium/nitrogen separation. The computational approach not only pinpointed the top-performing frameworks but also established structure–performance relationships, such as optimal pore-limiting diameters and accessible surface areas, which are critical for efficient separation. This methodology exemplifies how computational tools can streamline the design and selection of materials for industrial applications, reducing reliance on costly and time-consuming experimental trials.

This Special Issue highlights how first-principles and molecular dynamics simulations are transforming energy research across scales—from electronic structure and spin dynamics to interfacial phenomena and bulk material performance. Each contribution demonstrates the value of atomistic insights in guiding experimental design, improving material functionality, and enabling innovation in sustainable energy technologies. As computational power and theoretical models continue to evolve, molecular simulation will remain a cornerstone of next-generation material science and molecular engineering.

Funding: This research received no external funding.

Conflicts of Interest: The authors declare no conflict of interest.

Abbreviations

The following abbreviations are used in this manuscript:

DFT	Density Functional Theory
TDDFT	Time-Dependent DFT
QC	Quantum Chemical
CFF	Classical Force-Field
GO	Geometry Optimization
MC	Monte Carlo
GCMC	Grand Canonical Monte Carlo
MD	Molecular Dynamics
SAGD	Steam-Assisted Gravity Drainage
HPAM	Hydrolyzed Polyacrylamide
MOF	Metal–Organic Frameworks

List of Contributions:

1. Akbar, H.; Abdur, R.; Ejaz, A.; Muhammad, S.K.; Shabeer, A.M.; Joonkyung, J. Modulating Optoelectronic and Elastic Properties of Anatase TiO_2 for Photoelectrochemical Water Splitting. *Molecules* **2023**, *28*, 3252. https://doi.org/10.3390/molecules28073252.
2. Oguz, I.C.; Jaouen, F.; Mineva, T. Exploring Spin Distribution and Electronic Properties in FeN_4-Graphene Catalysts with Edge Terminations. *Molecules* **2024**, *29*, 479. https://doi.org/10.3390/molecules29020479.
3. Rosen, B.; Sohlberg, K. The Stability of a Mixed-Phase Barium Cerium Iron Oxide under Reducing Conditions in the Presence of Hydrogen. *Molecules* **2023**, *28*, 1429. https://doi.org/10.3390/molecules28031429.
4. Manian, A.; Chen, Z.; Hudson, R.J.; Russo, S.P. Simulation of Solvatochromic Phenomena in Xanthione Using Explicit Solvent Methods. *Molecules* **2024**, *29*, 5609. https://doi.org/10.3390/molecules29235609.
5. Liu, Z.; Hanna, G. Population and Energy Transfer Dynamics in an Open Excitonic Quantum Battery. *Molecules* **2024**, *29*, 889. https://doi.org/10.3390/molecules29040889.
6. Ahmadi, M.; Chen, Z. Molecular Dynamics Investigation of Wettability Alteration of Quartz Surface Under Thermal Recovery Processes. *Molecules* **2023**, *28*, 1162. https://doi.org/10.3390/molecules28031162.
7. Hue, K.Y.; Lew, J.H.; Matar, O.K.; Luckham, P.F.; Muller, E.A. Parametric Studies of Polyacrylamide Adsorption on Calcite Using Molecular Dynamics Simulation. *Molecules* **2025**, *30*, 285. https://doi.org/10.3390/molecules30020285.
8. Grenev, I.V.; Gavrilov, V.Y. In Silico Screening of Metal-Organic Frameworks and Zeolites for He/N_2 Separation. *Molecules* **2023**, *28*, 20. https://doi.org/10.3390/molecules28010020.

References

1. Chu, S.; Majumdar, A. Opportunities and Challenges for a Sustainable Energy Future. *Nature* **2012**, *488*, 294–303. [CrossRef] [PubMed]
2. Armaroli, N.; Balzani, V. *Energy for a Sustainable World: From the Oil Age to a Sun-Powered Future*; Wiley-VCH: Weinheim, Germany, 2011. [CrossRef]
3. Schmauder, S.; Schäfer, I. (Eds.) *Multiscale Materials Modeling: Approaches to Full Multiscaling*; De Gruyter: Berlin, Germany; Boston, MA, USA, 2016. [CrossRef]
4. Jun, H.K.; Low, F.W. (Eds.) *Materials for Energy Conversion and Storage*; CRC Press: Boca Raton, FL, USA, 2024. [CrossRef]
5. Walsh, A.; Sokol, A.A.; Catlow, C.R.A. (Eds.) *Computational Approaches to Energy Materials*; John Wiley & Sons, Ltd.: Hoboken, NJ, USA, 2013. [CrossRef]
6. Wang, L. (Ed.) *Molecular Dynamics—Theoretical Developments and Applications in Nanotechnology and Energy*; InTech: Vienna, Austria, 2012. [CrossRef]
7. Kang, J.; Zhang, X.; Wei, S.-H. Advances and challenges in DFT-based energy materials design. *Chin. Phys. B* **2022**, *31*, 107105. [CrossRef]

8. Jin, J.; Pak, A.J.; Durumeric, A.E.P.; Loose, T.D.; Voth, G.A. Bottom-up Coarse-Graining: Principles and Perspectives. *J. Chem. Theory Comput.* **2022**, *18*, 5759–5791. [CrossRef] [PubMed]
9. Wehinger, G.D.; Ambrosetti, M.; Cheula, R.; Ding, Z.-B.; Isoz, M.; Kreitz, B.; Kuhlmann, K.; Kutscherauer, M.; Niyogi, K.; Poissonnier, J.; et al. Quo vadis multiscale modeling in reaction engineering?—A perspective. *Chem. Eng. Res. Des.* **2022**, *184*, 39. [CrossRef]
10. Fish, J.; Wagner, G.J.; Keten, S. Mesoscopic and multiscale modelling in materials. *Nat. Mater.* **2021**, *20*, 774–786. [CrossRef] [PubMed]

molecules

MDPI

Article

Parametric Studies of Polyacrylamide Adsorption on Calcite Using Molecular Dynamics Simulation

Keat Yung Hue, Jin Hau Lew, Omar K. Matar, Paul F. Luckham and Erich A. Müller *

Department of Chemical Engineering, Imperial College London, London SW7 2AZ, UK; v.hue20@imperial.ac.uk (K.Y.H.); s.lew20@imperial.ac.uk (J.H.L.); o.matar@imperial.ac.uk (O.K.M.); p.luckham01@imperial.ac.uk (P.F.L.)

* Correspondence: e.muller@imperial.ac.uk

Abstract: This study investigates the efficacy of polyacrylamide-based polymers, specifically hydrolysed polyacrylamide (HPAM), in reducing solids production within carbonate reservoirs. Building on our earlier simulation approach, molecular simulations were conducted to examine how these polymers adsorb onto calcite, the main mineral found in carbonate formations. The adsorption process was affected by several factors, including polymer molecular weight, charge density, temperature, and salinity. Generally, increased molecular weight, charge density, and temperature resulted in higher adsorption rates. The effect of salinity was more nuanced, as salt-bridging and charge-screening effects created competing influences. The simulation outcomes correspond closely with experimental results, offering valuable insights for designing and optimizing polymer-based strategies aimed at controlling solids production in carbonate reservoirs.

Keywords: adsorption; molecular simulation; parametric studies; surfaces; polymers

1. Introduction

In the energy industry, the term "formation strength" refers to the mechanical properties of reservoir rocks to withstand the stresses exerted upon them. Solids production becomes an issue in weak and poorly consolidated carbonate formations when the induced formation stress surpasses the formation strength [1–3]. This can lead to the detachment of solid particles which are then transported with the reservoir fluids, potentially wearing surface equipment and increasing environmental waste [4]. The solids production risk can be reduced by injecting formation-strengthening chemicals, where these chemicals have been investigated experimentally [5]. However, screening suitable chemicals can be time-consuming and experimentally challenging. Molecular simulation offers an alternative approach to expedite chemical-screening processes by suggesting potential candidates to complement experimental synthesis. The performance of these chemicals can be evaluated in terms of their molecular conformation and adsorption characteristics onto calcite surfaces, which is the main component of carbonate formations.

In our previous manuscript [6], various polyacrylamide-based (PAM) polymers, including basic PAM, hydrolysed polyacrylamide (HPAM), and sulfonated polyacrylamide (SPAM), were evaluated using adsorption free energy analysis at room temperature and in vacuum conditions, with HPAM showing the best adsorption performance. However, polymer adsorption may be influenced by more challenging reservoir environments characterised by higher pressure, temperature, and salinity. Additionally, polymer characteristics such as molecular weight and charge density have been shown to impact polymer adsorption mechanisms. Understanding the effects of these parameters on polymer–calcite interactions is essential for predicting polymer adsorption performance to complement experimental studies.

It is also important to note that polymers are typically injected as solutions during formation consolidation, requiring parametric studies to be conducted in aqueous environments. Nevertheless, previous work has shown that in water solvents, the highly hydrophilic PAM-based polymers result in nearly zero free energy value, complicating the analysis of polymer performance across different parametric studies. This manuscript will deploy the equilibrium adsorption simulation approach, focusing on polymer adsorption density on calcite surfaces in aqueous environments under varying polymer chemistries and reservoir conditions.

Literature Review of Adsorption Parametric Studies

Extensive experimental and simulation studies have been conducted to investigate the effects of various parameters on polymer dynamics and their adsorption behaviours on mineral surfaces. These parameters include polymer molecular weight [7–13], charge density [7,8,12,14], functional group [13,15–22], pH [12,23,24], temperature [11,17,25–28], pressure [11,17], and salinity conditions [10,11,17,18,23,25,29–35]. Although not limited to formation-strengthening applications, MD simulations of PAM-based polymers have been predominantly reported for wastewater treatment and enhanced oil recovery (EOR) applications. This can be attributed to their excellent viscosifying and flocculating properties and their capability to form crosslinked hydrogels [36–38].

Abdel-Azeim and Kanj [11] conducted parametric studies on the dynamic and aggregation behaviour of HPAM polymer in aqueous solution. They found that increasing temperature and pressure did not significantly affect the polymer's radius of gyration, R_g, and end-to-end distance, R_e. While increasing the polymer's molecular weight results in a higher R_g, there is no consistent trend in polymer conformation, which can range diversely from coiled to extended shapes. In salt-free conditions, the polymer adopts an extended conformation due to electrostatic repulsion among the negatively charged functional groups. In contrast, a high salinity environment neutralises these charges, leading to a transition to a coiled conformation with reduced R_g. Chen et al. and Quezada et al. reported similar effects of salinity on PAM polymer conformational behaviour [31], with Quezada et al. concluding that this impact strongly correlated with polymer charge density. In their study of polymers, including PAM, HPAM, and SPAM in high-salinity solutions, Quezada et al. found that HPAM coiling was the most intense, followed by SPAM and PAM. This is because PAM is a neutral molecule, and SPAM has a lower charge strength and ion adsorption capacity compared to HPAM. Consequently, HPAM exhibits a lower R_g, potentially limiting its effectiveness in forming polymeric bridges for particle flocculation under high-salinity conditions.

The literature on adsorption parametric studies of PAM-based polymers on calcite surfaces is sparse. Ahsani et al. [28] employed both experimental studies and MD simulations to investigate the wettability alteration behaviour of HPAM on calcite surfaces. While calcite surfaces have an ionic structure and are inherently water-wet, natural carbonate formations are often surrounded by hydrocarbons with negatively charged carboxylates, which anchor to the surface and render them oil-wet. Their experiments recreated this calcite surface condition with pre-adsorbed oil components containing carboxylates. They found that HPAM could form a polymeric layer with surfactant-like behaviours to interact with the carboxylates or competitively adsorb to the surface and desorb the carboxylates, thereby modifying the calcite surface to a more water-wet state. A simpler model was used in their MD simulation without considering the pre-adsorbed carboxylates. The contact angle of water indicated HPAM adsorption based on the surface tension of the water–

polymer–calcite phases. They observed that HPAM adsorption improved with increasing temperature from 20 °C to 80 °C in experiments, but a contradictory behaviour was noted in simulations. This disagreement in simulation results is attributed to increased polymer desorption with increasing kinetic energy. Another possible reason for the discrepancy could be the calcite crystal plane used in the simulations, where the surface consisted only of the carbonate (CO_3^{2-}) layer, resulting in stronger electrostatic repulsion with HPAM.

While not focusing on PAM polymers, Chen et al. [29] studied the interactions of a carbohydrate molecule, glucopyranose, with different calcite crystal planes. They observed that the adsorption of the carbohydrate decreased with increasing brine concentration across all crystal planes. Since the carbohydrate is a neutral molecule, the primary mechanism of calcite–carbohydrate interaction is governed by hydrogen bonding, with the crystal plane having the highest oxygen density showing the strongest affinity to the hydrogen atoms from the carbohydrates. Increasing brine concentration contributes to the formation of salt layers on the calcite surface, which screens the hydrogen bonding between calcite and carbohydrates and inhibits adsorption. They inferred that charged polymers typically used in EOR applications in high-salinity environments might follow a different mechanism. The charged functional groups of these polymers can have favourable electrostatic interactions with the adsorbed cations, resulting in increased polymer retention via salt bridging. Thus, they suggested that alternative non-ionic polymers could reduce polymer damage and retention on calcite surfaces.

Meanwhile, there are more simulation studies on the adsorption of PAM-based polymers on montmorillonite and quartz, both containing negatively charged surfaces. Ma et al. [16] investigated PAM-based polymers with different ionicities and found that cationic PAM had the most substantial adsorption to montmorillonite, followed by neutral and anionic PAM. This can be attributed to the electrostatic attraction from the cationic functional group, $N(CH_3)^{3+}$, and hydrogen bonding from the amide functional group, $CONH_2$. A similar effect of polymer ionicity was reported by Qu et al. [17], who investigated different types of polymers, including neutral PAM. Additionally, they observed that higher pressure had a negligible effect on adsorption, and higher temperature reduced the adsorption of all polymers except PAM, whose adsorption remained stable at 300 K and 400 K with slight desorption at 500 K. Interestingly, despite the negatively charged surface, increasing salinity can improve the adsorption of both cationic and anionic polymers with negligible influence on neutral polymers. Sun et al. [35] investigated the effect of cation valency and found that divalent Ca^{2+} ions enhanced the adsorption of anionic PAM to montmorillonite more effectively than monovalent Na^+ ions due to a stronger charge-screening effect and a reduction in electrostatic repulsion.

Regarding PAM adsorption on quartz surfaces, one of the most relevant studies for sand consolidation applications was conducted by Huang et al. [20]. They screened different types of potential chemicals and concluded that PAM exhibited the best adsorption performance. This performance was correlated to the strength of hydrogen bonding with the quartz surface, and the trend agreed with the reported experimental consolidation strength. In contrast, Yang et al.'s study [13] aimed to avoid polymer retention on quartz surfaces and assessed the performance of PAM-based polymers under different molecular weights and temperatures. They found that PAM adsorption increased with increasing molecular weight but decreased significantly with increasing temperature. Compared to Qu et al.'s study [17], the differing temperature trends of PAM adsorption on the montmorillonite surface suggest that interfacial properties may affect the adsorption mechanism. In another study, Quezada et al. [23,30] investigated the adsorption of HPAM on quartz under high-salinity conditions. They found that cations neutralised the electrostatic repulsion between HPAM and quartz, thereby improving adsorption. This finding is consistent with the results for the montmorillonite case.

To summarise, various parametric studies have been reported on the adsorption of PAM-based polymers on different surfaces. However, comprehensive research has not yet been explored to address the combined effects of reservoir conditions and polymer chemistry on adsorption to calcite surfaces, tailored explicitly for formation-strengthening applications. This gap underscores the need for further investigation. In this manuscript, we examine the influence of polymer molecular weight, charge density, temperature, pressure, and salinity on adsorption to calcite surfaces at typical reservoir conditions, with simulation results compared to experimental data whenever available.

2. Results and Discussion

2.1. Effect of Polymer Molecular Weight

The polymer adsorption, measured through both simulations and experiments at different molecular weights (MW), is depicted in Table 1, along with the simulated R_g and κ^2. In simulations, polymer MW was represented in terms of repeat units (213 Da per repeat unit) as a surrogate for experimental candidates with varying MW. Despite experimental polymers having significantly higher MW, potentially reaching millions of Daltons, there was good agreement between simulations and experiments, with increasing polymer adsorption observed as molecular weight increased. This trend was expected because longer molecular chains experience stronger interactions with the calcite surface, increasing the contact area between the polymer and the surface and enhancing adsorption strength. In addition, since HPAM is a hydrophilic polymer, the shortest HPAM chains prefer to remain in the water and require a longer time to stabilise their adsorption onto the calcite surface.

Table 1. Polymer adsorption comparison between experiments and MD simulations, R_g and κ^2 at different molecular weights.

HPAM Repeat Units	Experimental Adsorbed Amount (mg/m^2)	MD Adsorbed Amount (1/nm^2)	R_g (Å)	κ^2
10	0.250 ± 0.014	2.731 ± 0.575	9.48 ± 1.21	0.474 ± 0.111
20	0.330 ± 0.020	3.476 ± 0.370	10.27 ± 0.38	0.211 ± 0.068
30	0.420 ± 0.010	4.370 ± 0.126	11.97 ± 0.27	0.126 ± 0.032

While polymer MW positively impacts adsorption, both the experimental and simulation literature indicate that adsorption will reach an equilibrium value. Once the polymer adsorption reaches saturation, further increases in polymer MW or concentration do not enhance adsorption [7,8,39] and may even reduce stability [13]. This is attributed to changes in the adsorption mechanism with increasing polymer MW. Experimentally, it has been observed that polymer adsorption via polymer bridging is the strongest mechanism, followed by electrostatic bare patching [40]. At low MW, polymers primarily adopt a flat, extended conformation on the calcite surface, with electrostatic bare patching being the dominant adsorption mechanism. As MW or concentration increases, the polymer chains become compressed with a reduced effective contact area for each chain. This leads to the formation of more loops, with polymer bridging becoming the predominant adsorption mechanism [7,8,39]. Further increases in MW do not enhance adsorption due to the limited contact area and may inhibit existing polymer bridging between particles, potentially affecting adsorption stability.

In our simulation, given that only a single polymer chain was modelled and the calcite surface area was not saturated with polymer adsorption, the equilibrium adsorbed amount was not observed but was expected to be reached with higher polymer repeat units. Nevertheless, the polymer conformation with different MWs can be characterised by κ^2, as shown in Table 1. The increasing R_g with higher MW was expected due to the larger polymer size. In contrast, the decrease in κ^2 with increasing MW indicates that the polymer transitions from an extended conformation to a coiled conformation, showing excellent agreement with theoretical expectations.

Since both the polymer adsorbed amount and R_g increase with increasing MW, it is interesting to determine if a proportional relationship between them can be established in this study. We propose a simple relation to relate polymer R_g with the polymer repeat unit. Polymer R_g can be considered as the radius of the polymer globule, where the globular volume, V, can be approximated as the volume of a sphere:

$$V \approx \frac{4}{3}\pi {R_g}^3 \tag{1}$$

The total polymer mass, M, is directly proportional to the number of repeat units, N, and the mass of a single repeat unit, m:

$$M = Nm \tag{2}$$

Assuming the polymer density, ρ, is constant, the polymer mass can be expressed in terms of V and ρ, approximated as shown in Equation (3). The number of repeat units thus has a proportional relationship with the R_g as indicated in Equation (4).

$$Nm = \rho V \approx \rho \frac{4}{3}\pi {R_g}^3 \tag{3}$$

$$\sqrt[3]{N} \propto R_g \tag{4}$$

This proportionality was investigated by dividing the polymer R_g and number of adsorbed atoms using the normalization factor $\sqrt[3]{N}$, with the results presented in Figure 1. To further validate this proportionality, another case of HPAM with 60 repeat units was also modelled in a larger water–calcite system to avoid system size effects. After normalization, it could be observed that the polymer R_g and adsorbed amount reached a constant range with varying polymer repeat units, suggesting that their values increased proportionally with the cubic root of polymer repeat units during adsorption.

This implies that the R_g of the polymer scales with MW at an exponent of 0.33 during adsorption to calcite surfaces, which diverges from the expected scaling relationship of greater than 0.5 typically seen in free solution under good solvent conditions [41]. This discrepancy can be attributed to the conformational changes that occur when HPAM interacts with the surface. Upon adsorption, the HPAM exhibits reduced flexibility, transitioning to a more coiled or globular conformation that is influenced by the increasing MW (as observed in $\kappa^2 < 0.5$, indicating nearly spherical shape). The observed lower exponent is consistent with findings in the literature, which suggest that globular polymers demonstrate a scaling behaviour characterized by lower exponents [41]. This indicates that the adsorption environment significantly alters the polymer conformation, resulting in a unique scaling relationship that highlights the complexities of polymer behaviour at interfaces.

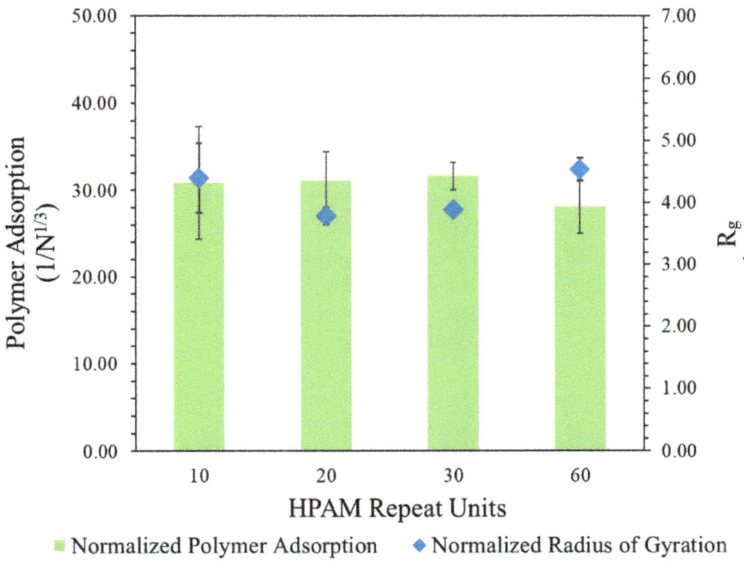

Figure 1. Polymer adsorption and radius of gyration, R_g, with normalization factor of cubic root of number of repeat units, $\sqrt[3]{N}$, at different molecular weights. Green bar indicates normalized polymer adsorption (left axis) and blue diamond symbol indicates normalized R_g (right axis), with error bars representing standard error averaged over 5 realizations. Error bar of R_g is significantly smaller, except for HPAM with 10 repeat units.

2.2. *Effect of Charge Density*

Charge density (CD) is another polymer characteristic that can be fine-tuned to improve adsorption to calcite surfaces. The HPAM adsorption measured by simulations and experiments at different CDs and the polymer R_gs and κ^2s are detailed in Table 2. As CD increased, polymer adsorption improved with an agreeable trend in experiments and simulations. This improvement is attributed to the presence of more deprotonated carboxyl groups from the acrylate groups in the polymer (–COO^-), which exhibit stronger attractive electrostatic interactions with the calcite surface.

Table 2. Polymer adsorption comparison between experiments and MD simulations, R_g and κ^2 at different charge densities.

HPAM Charge Density (%)	Experimental Adsorbed Amount (mg/m^2)	MD Adsorbed Amount (1/nm^2)	R_g (Å)	κ^2
10	0.160 ± 0.008	2.634 ± 0.386	10.01 ± 0.19	0.151 ± 0.028
30	0.240 ± 0.014	3.596 ± 0.649	11.55 ± 0.93	0.276 ± 0.095
40	0.265 ± 0.011	4.580 ± 0.563	13.41 ± 1.33	0.370 ± 0.094

Since all polymer models have similar repeat units and sizes, the trends in both R_g and κ^2 are identical. The increase in both R_g and κ^2 with increasing CD indicates that the polymer transitions to a more extended conformation during adsorption. This is due to the increasing electrostatic repulsions from the negatively charged functional groups in the chain, causing polymer segments to push away from each other and resulting in a more linear chain [42]. Meanwhile, a polymer with lower CD experiences less repulsion among the segments and prefers a coiled, more compact structure. The adsorption results reported here agree with the theory, with the literature suggesting that an optimum polymer CD should be employed as overly charged polymers can trigger charge reversal on the particle surfaces, inhibiting polymer bridging and adsorption [32,40].

2.3. Effect of Temperature

The results of polymer adsorption under the influence of temperature are interesting. As shown in Table 3, experimental data indicate that polymer adsorption deteriorates with increasing temperature, whereas simulations show an opposite trend. Theoretically, polymer adsorption is generally exothermic, and increasing temperature shifts the equilibrium state towards desorption [25]. This effect is further exacerbated by polymer degradation at high temperatures, causing the polymer to lose its effectiveness in bridging particles. The experimental results align with this theory, indicating that the adsorption effectiveness of HPAM could be inhibited by increasing temperature.

Table 3. Comparison of polymer adsorption between experiments and simulations at different temperatures.

System Temperature (°C)	Experimental Adsorbed Amount (mg/m^2)	MD Adsorbed Amount (1/nm^2)
25	0.240 ± 0.014	4.370 ± 0.126
60	0.190 ± 0.018	6.642 ± 1.158
100	0.110 ± 0.011	9.052 ± 0.915

Meanwhile, the discrepancy in simulations with increasing polymer adsorption under higher temperature, though interesting, is not uncommon and has been reported in other studies. For example, a literature study on molecular adsorption on liposome surfaces observed rapid molecular transport kinetics at high temperatures with more negative adsorption free energy [43]. At higher temperatures, liposomes experience increased kinetic energy, which can influence adsorption behaviour depending on the enthalpic and entropic contributions to Gibbs free energy. Although adsorption typically results in decreased entropy due to reduced disorder, the author noticed a positive change in entropy. This is attributed to the replacement of water and counterions from the surface by adsorbate molecules, leading to an overall increase in entropy upon adsorption. To investigate the transport properties of HPAM, polymer mean squared displacement, MSD, and self-diffusion coefficients, D, are computed below [44]:

$$MSD = \left\langle [\boldsymbol{r}_i(t) - \boldsymbol{r}_i(0)]^2 \right\rangle \tag{5}$$

$$D = \frac{1}{6} \lim_{t \to \infty} \frac{d}{dt} \left\langle [\boldsymbol{r}_i(t) - \boldsymbol{r}_i(0)]^2 \right\rangle \tag{6}$$

where $\boldsymbol{r}_i(0)$ is the coordinate at the initial time and $\boldsymbol{r}_i(t)$ is the coordinate at different times. D can be measured from the slope fitted over the linear region of the MSD versus the time plot, as depicted in Figure 2 for polymer MSD at different temperatures. It is important to note that D is typically valid only for free polymers in solution. In this study, D is computed to characterise the polymer transport behaviour during adsorption and is measured over the linear slope at the initial timesteps. The computed D, along with polymer R_g and κ^2, are presented in Table 4.

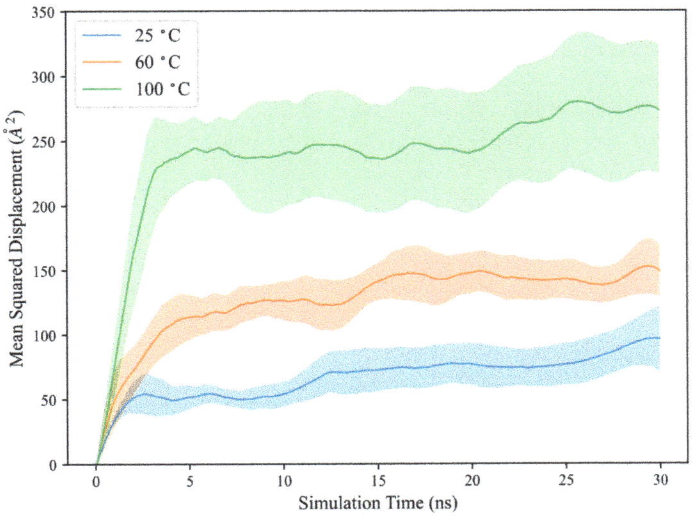

Figure 2. Polymer mean squared displacement at different temperatures. Shaded areas represent standard error from 5 realizations.

Table 4. Polymer R_g, κ^2, and D at different temperatures.

System Temperature (°C)	R_g (Å)	κ^2	D (10^{-11} m^2/s)
25	11.97 ± 0.27	0.126 ± 0.032	4.30 ± 1.08
60	12.51 ± 0.36	0.132 ± 0.024	5.88 ± 1.39
100	14.19 ± 0.95	0.268 ± 0.068	13.85 ± 3.71

The results indicate that polymer MSD and D are significantly higher at elevated temperatures, suggesting that the polymer could diffuse more rapidly towards the calcite surface with increasing temperatures. The polymer could more vigorously explore its surrounding configurational space, leading to an increased contact area with the calcite surface and enhanced adsorption. This is further supported by the increase in R_g and κ^2 with temperature, signifying that the polymer could adopt a more extended conformation to maximize adsorption contact with the calcite surface. Additionally, solvent quality may change with temperature, which could also influence the polymer–calcite interaction, although this is not considered in the current model.

In the literature, the influence of temperature on polymer adsorption has shown different trends in both experiments and simulations. Some studies report that polymers experienced enhanced adsorption with increasing temperatures [27,28,43,45], while others indicate increased desorption with higher temperatures [13,17,25]. The polymer–surface interactions can further complicate this inconsistency and require careful examination. In addition, a limitation of the simulation approach is its inability to account for polymer degradation at high temperatures. Our simulations do not consider structural changes in the polymer, which could be modelled only with a reactive forcefield. Consequently, the observed discrepancy between the simulation and experimental results may be attributed to the simulation focusing solely on thermodynamic factors without incorporating polymer degradation effects.

2.4. Effect of Pressure

As no experimental results are available for polymer adsorption under varying pressure, only simulation results will be discussed here. These results, including polymer adsorption, R_g, and κ^2 are summarised in Table 5. Increasing pressure showed no significant changes and led to only slight increases in polymer adsorption, R_g and κ^2. This finding is consistent with the existing literature, which indicates that pressure has a negligible effect on adsorption and does not significantly impact polymer conformational behaviour [17]. This is expected, as the polymer solution remains in a liquid state and is less affected by pressure. Nevertheless, it has been reported that the rheological properties of PAM solutions, such as viscosity, can improve under extremely high pressures (above 200 bar) [46]. PAM exhibits noticeable shear-thickening behaviour at high-shear-stress deformations due to stronger intermolecular interactions and the formation of transient networks between polymer chains. While the enhanced viscosity of PAM under high pressure may strengthen polymer bridging, this effect is not captured in our adsorption study and requires further investigation to assess the impact of high pressure on PAM solution viscosity.

Table 5. Polymer adsorption, R_g, and κ^2 at different pressures.

System Pressure (atm)	MD Adsorbed Amount ($1/nm^2$)	R_g (Å)	κ^2
1	4.525 ± 0.561	11.91 ± 0.25	0.123 ± 0.023
80	5.438 ± 0.385	11.91 ± 0.23	0.106 ± 0.019
160	5.237 ± 0.181	12.06 ± 0.36	0.123 ± 0.022

2.5. Effect of Salinity

In this section, the adsorption of various PAM-based polymers, including HPAM 33%, SPAM 33%, and neutral PAM, are investigated under varying cation valencies and ion sizes to observe the effect of their functional groups in saline environments, as presented in Figure 3a. The cation adsorption density on the surface is analysed similarly to polymer adsorption, with results displayed in Figure 3b. Additionally, radial distribution analysis is conducted to examine the adsorption of cations onto the polymer, as shown in Figure 3c. This is computed via the coordination number of the cations with respect to the carbonyl oxygen atom of the polymer amide group ($C(=O)NH_2$). Lastly, the polymer conformational size is analysed using R_g in Figure 3d.

The polymer adsorption results show that HPAM generally performs better than SPAM 33% and neutral PAM, with relatively high adsorption amounts measured across all cation cases. As explained in the previous adsorption free energy analysis, this can be attributed to the presence of carboxyl functional groups in HPAM, which experience stronger electrostatic attraction with the calcite surface compared to the bulkier functional groups in SPAM and the neutral ionicity of PAM. When comparing monovalent cations (Li^+, Na^+, K^+) and divalent cations (Mg^{2+}, Ca^{2+}, Sr^{2+}), interesting trends emerge among different polymers. HPAM adsorption is higher in monovalent cation cases than in divalent cation cases, with increasing monovalent cation sizes leading to increased adsorption, while increasing divalent cation sizes result in decreased adsorption. SPAM also shows reduced adsorption in divalent cation cases compared to monovalent cations. However, unlike HPAM, increasing monovalent cation sizes leads to decreased adsorption, while increasing divalent cation sizes results in increased adsorption. On the other hand, PAM adsorption is less influenced by cations. PAM shows better adsorption in divalent cation cases than in monovalent cation cases, with increasing monovalent cation sizes resulting in decreased adsorption and no significant trend observed for increasing divalent cation sizes.

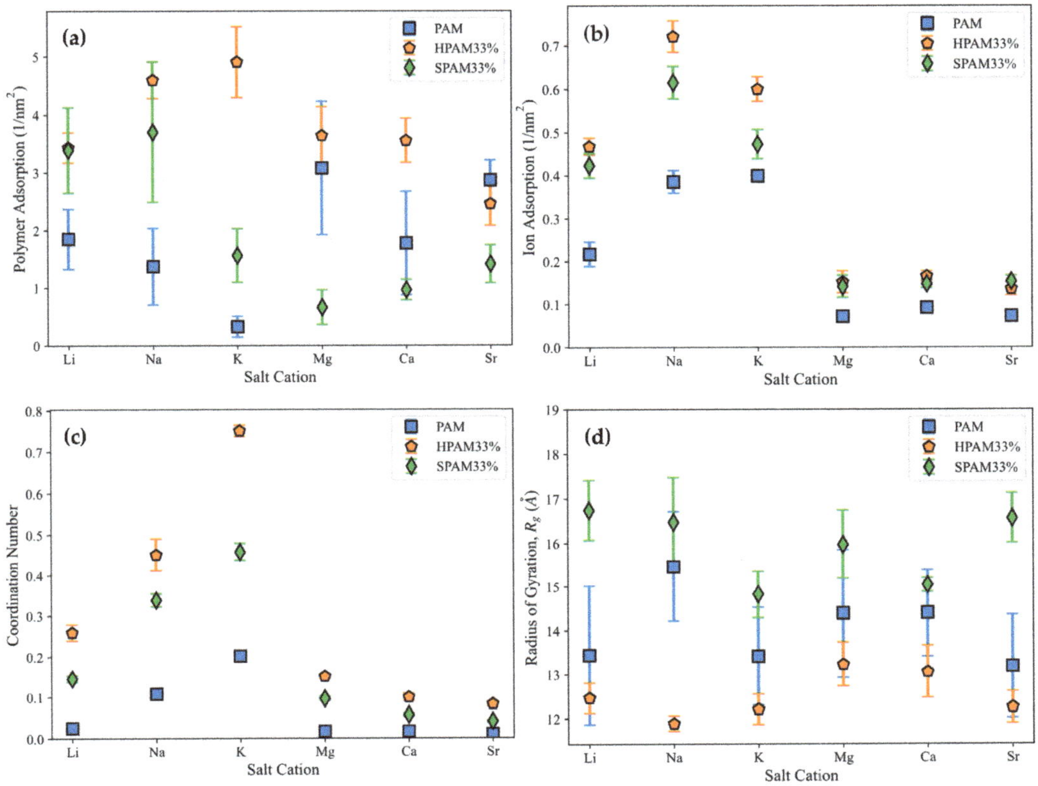

Figure 3. Simulation results of polymer adsorption with different salt cations: (**a**) Polymer adsorption to calcite surface, (**b**) cation adsorption to calcite surface, (**c**) cation adsorption to polymer, (**d**) polymer R_g. Polymer or cation adsorption to calcite surface are defined as the number of polymer chain atoms or cation atoms normalized by the surface area, counted within a layer thickness of 4 Å above the surface. Cation adsorption to polymer is defined as coordination number of cations with respect to carbonyl oxygen atom of polymer amide group ($C(=O)NH_2$) in radial distribution function analysis. Blue square, orange pentagon, and green diamond indicate PAM, HPAM 33%, and SPAM 33%, respectively, with error bars representing standard error from 5 realizations.

The differing adsorption behaviours among various polymers can be attributed to their distinct adsorption mechanisms influenced by cations. In the literature, the roles of cations in enhancing or inhibiting polymer adsorption have yielded contradictory results, with different mechanisms proposed to explain these behaviours. For polymer adsorption enhancement, some authors suggest that salt ions can form an Electric Double Layer (EDL) on the surface, facilitating salt bridging that attracts the polymer to the surface regardless of the polymer's ionicities [17,23]. Conversely, polymer inhibition is attributed to the charge-screening effect of ions, which suppresses the polymer charge or causes surface charge reversal, thereby reducing the polymer's adsorption strength to the surface [11,35]. In a saline environment, the adsorption of PAM-based polymers onto the calcite surface, focusing on HPAM, can be illustrated with the schematic diagram in Figure 4, with an electrostatic surface potential diagram provided as Figure S2 in the Supplementary Information.

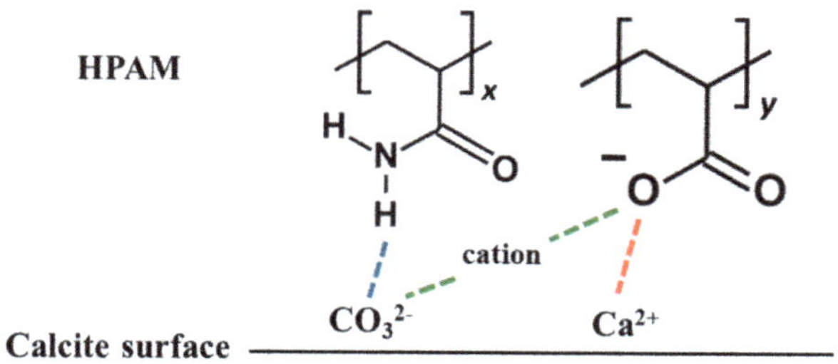

Figure 4. Schematic diagram of possible adsorption mechanism. **Blue bond**: hydrogen bonding between carbonate ion and amide group; **red bond**: charge interaction between calcium ion and carboxyl group; and **green bond**: salt bridging between carbonate anion and carboxyl group.

Effective polymer adsorption to the calcite surface results from the competing effects of salt bridging and charge screening. In experiments, the calcite surface is positively charged under normal reservoir pH conditions [47], where cations may weaken the charge of an anionic polymer and screen its attraction to the surface. However, the calcite surface in the simulation is neutral, consisting of alternating CO_3^{2-} and Ca^{2+} ions. HPAM can form hydrogen bonds with calcite CO_3^{2-} via hydrogen atoms from the amide groups or significantly stronger charge interactions between calcite Ca^{2+} and deprotonated oxygen atoms from the carboxyl groups. While cations may neutralise the charge attraction between Ca^{2+} and deprotonated carboxyl groups, they can also function as salt bridges between CO_3^{2-} and carboxyl groups, enhancing adsorption.

Simulation results suggest that salt bridging is more dominant in HPAM adsorption. HPAM's higher charge density makes it more susceptible to cation influence than SPAM and PAM [31]. This claim agrees with the density profile analysis of the relevant oxygen atoms of the polymers and the cation (using Na^+ as an example) during adsorption, as shown in Figure S3 in the Supplementary Information. This enhances the role of salt bridging, contributing to its higher adsorption across various cation cases. The differing adsorption trends between monovalent and divalent cations can be attributed to the hydration energy of the cations. Cations with stronger charge density form a more robust hydrogen-bonding network with water, exhibiting stronger hydration energy and preferring to remain in the bulk water phase. Such cations are known as water structure makers, while those with weaker hydration energy are termed water structure breakers [23].

Divalent cations exhibit significantly higher charge density than monovalent cations, while increasing ion sizes within the same valency can reduce charge density. Consequently, the hydration energy follows the following trend: $Mg^{2+} > Ca^{2+} > Sr^{2+} > Li^+ > Na^+ > K^+$ [48]. This trend is reflected in the cation adsorption density shown in Figure 3b, where divalent cations, with stronger hydration energy, exhibit lower surface adsorption and prefer to remain in the aqueous phase compared to monovalent cations. The trend observed for monovalent cations also aligns with the hydration energy trend. In HPAM and SPAM cases, the lower adsorption of K^+ compared to Na^+ may be attributed to its larger ionic radius, which positions the adsorption peak further from the surface and may extend beyond the defined 4 Å adsorption layer [49].

The variation in cation adsorption density on the surface across different polymers indicates the influence of the polymer on cation adsorption. On the other hand, cation adsorption to the surface also affects the density of cations adsorbed onto the polymer. Higher cation adsorption to the surface creates a more concentrated salt layer, which attracts the polymer more effectively and increases the cation adsorption onto the polymer. As shown in Figure 3c, the cation adsorption onto the polymer exhibits a trend similar to that in Figure 3b, highlighting the interplay between cations and polymers. Consequently, these two factors contribute to the adsorption trends observed for different polymers under varying cation conditions depicted in Figure 3a.

Since the adsorption of HPAM to the surface is primarily driven by salt bridging, the observed increase in adsorption with larger monovalent cations is due to stronger cation adsorption onto the polymer ($K^+ > Na^+ > Li^+$). Conversely, the decrease in adsorption with larger divalent cations is attributed to the reduced cation adsorption density on the polymer ($Mg^{2+} > Ca^{2+} > Sr^{2+}$). The simulation snapshots in Figure 5 further illustrate this. Closer inspection reveals that a stable EDL is formed for monovalent cations, with cations creating the Stern layer (the first EDL layer) and Cl^- anions constituting the diffuse layer (the second EDL layer), stabilising the adsorption. In contrast, no stable EDL exists for divalent cations, resulting in lower adsorption. This observation aligns with previously reported cation interfacial interaction behaviours in calcite–water systems [49].

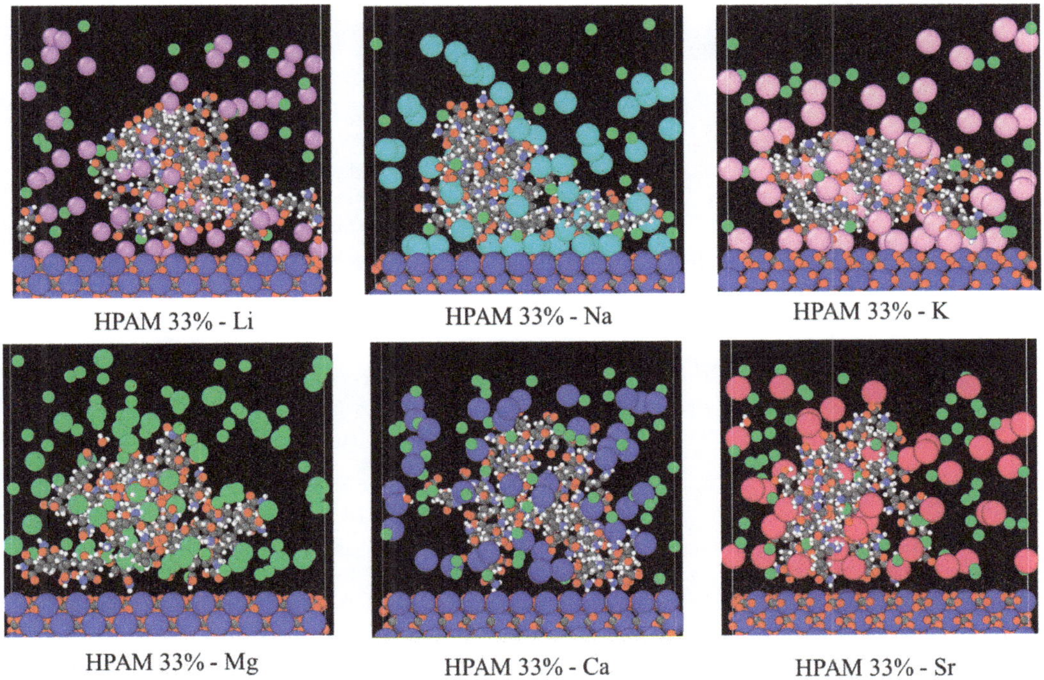

Figure 5. Simulation snapshot of HPAM 33% adsorption with different salt cations. Smaller particle—anions: Cl^- (forest green). Larger particle—cations: Li^+ (purple), Na^+ (cyan), K^+ (pink), Mg^{2+} (mint green), Ca^{2+} (dark blue), Sr^{2+} (magenta).

On the other hand, the differing trends observed for SPAM may be attributed to its lower charge strength, where cations have a reduced influence on polymer adsorption [31] or might even inhibit it through charge-screening effects. Compared to monovalent ions, SPAM exhibits lower adsorption to a calcite surface in divalent ions due to the stronger charge-screening effects. The decreasing adsorption trend with increasing monovalent cation size is attributed to stronger cation adsorption and greater charge screening. Conversely, the increasing adsorption trend with larger divalent cations results from a relatively lower charge-screening effect combined with reduced cation adsorption density.

Meanwhile, PAM, which has a neutral charge, primarily adsorbs onto the calcite surface through hydrogen bonding from amide groups, with adsorption governed mainly by surface oxygen densities [29]. Cations can inhibit this adsorption by screening the interaction between hydrogen atoms and calcite CO_3^{2-}. Consequently, the lower cation adsorption density on both the polymer and surface for divalent cations leads to higher PAM adsorption than monovalent cations. The decreasing trend in adsorption with increasing monovalent cation size is due to increasing cation adsorption density, resulting in more pronounced charge-screening effects. While cations inhibit adsorption in the case of PAM, they function as salt bridges between the anionic polymer and calcite CO_3^{2-}. For neutral PAM, anions may thus play a more dominant role in bridging interactions between the hydrogen atoms of the amide group and Ca^{2+}. Although not presented here, a published study investigated the role of anions, showing that nitrate ions (NO_3^-) with a smaller hydrated radius and hydration energy than Cl^- can enhance the adsorption of neutral PAM on the calcite surface [50].

As illustrated in Figure 3d, polymer R_g shows no significant trend across different cation cases. Among the polymers, SPAM has the largest size, followed by PAM and HPAM. Since all polymers have the same degree of polymerization, SPAM naturally exhibits the largest size due to its bulkier functional groups. Conversely, HPAM, with the smallest R_g, suggests a more compact structure. This is expected, as the cations strongly adsorb onto HPAM, neutralising its charge and causing the polymer to adopt a more compact conformation.

We conclude this section by comparing the experimental and simulation results for HPAM adsorption in salt-free conditions (control), NaCl, and $CaCl_2$ environments, as presented in Table 6. Both experimental and simulation results show a similar trend for salt-free and Na^+ cases, indicating that salt cations can enhance polymer adsorption through salt bridging. However, for divalent Ca^{2+}, the simulation shows lower adsorption compared to Na^+, while experiments reveal significantly higher adsorption. This discrepancy may be attributed to different interpretations of polymer adsorption. In experiments, Ca^{2+} provided more charges to neutralise the polymer, causing it to adopt a more coiled conformation and enabling more polymer chains to adsorb onto calcite particles, thereby increasing overall adsorption. In contrast, since only a single polymer chain is modelled in the simulation, the effects of multiple polymer chains adsorbing to the surface cannot be captured, leading to different observations.

Table 6. Comparison of polymer adsorption between experiments and simulations at salt cation valencies.

Salt Cation Valency	Experimental Adsorbed Amount (mg/m^2)	MD Adsorbed Amount ($1/nm^2$)
Salt-free condition	0.330 ± 0.020	4.370 ± 0.126
Na	0.380 ± 0.036	4.593 ± 0.313
Ca	0.820 ± 0.137	3.554 ± 0.383

3. Methodologies

3.1. General Simulation Details

The simulation details outlined here were applied uniformly across all MD simulations discussed in the previous sections. Classical atomistic molecular dynamics (MD) simulation was employed using Material Exploration and Design Analysis (Medea) simulation software v.3.5 [51] integrated with the LAMMPS module [52] with a built-in visualisation interface. All the molecules were described by the enhanced version of the all-atom Polymer Consistent Force Field (PCFF+) actively developed by MedeA (see Section S1 in Supplementary Information and ref. [6] for more forcefield parameter details and validation in our previous work), with a forcefield cut-off distance of 9.5 Å and the treatment of long-range electrostatic interaction with the particle–particle–particle Mesh (PPPM) method [53]. System minimisation was performed using the conjugate gradient method, and the simulations were conducted with the Velocity Verlet algorithm with a timestep of 1 fs. Periodic boundary conditions were applied in all Cartesian directions. Whenever an isobaric-isothermal ensemble (NPT) or a canonical ensemble (NVT) was performed, the system was controlled using the Nose–Hoover thermostat and barostat with the correction terms of Martyna, Tuckerman, and Klein [52,54] included in the equations of motion.

3.2. Calcite Model

The calcite structure was modelled to represent the carbonate rock. The unit cell was a rhombohedral crystal structure with a space group of $R\bar{3}c$. It had a dimension of $a = b = 4.980$ Å, $c = 17.192$ Å, and a plane angle of $\alpha = \beta = 90°$, $\gamma = 120°$. It was further cleaved into a crystal plane (1 0 4), which has proven to be the most thermodynamically stable structure and has the lowest surface free energy, as validated in our previous work [6]. The crystal plane was replicated as a 6-layer thickness structure with 720 $CaCO_3$ molecules and dimensions of $a = 48.86$ Å, $b = 49.76$ Å and $c = 17.84$ Å. This calcite structure was utilised across all the parametric studies.

3.3. Polymer Model

HPAM 33%, the polymer with the best adsorption performance from our previous work [6], was selected as the basis polymer model in the parametric studies. However, depending on the parameter of interest, other PAM-based polymers, such as basic PAM and SPAM 33%, were also modelled as polymer candidates. The oligomer repeat unit composition and copolymer ratio have been illustrated and detailed in Table 7. The "repeat unit" refers to the smallest representative structure forming a copolymerised PAM with a specific copolymer ratio. In the cases of HPAM 33% and SPAM 33%, the repeat unit comprised two acrylamide monomers and a deprotonated copolymer group, resulting in a structure with 33% charge density. In contrast, the corresponding size of the repeat unit of pure PAM consisted of three acrylamide monomers.

Table 7. Polymer compositions and molecular weight. *x* and *y* refer to the ratio of copolymers in a repeat unit.

Polymer	Repeat Unit Composition	Copolymer Ratio	Molecular Weight per Repeat Unit (Da)
PAM	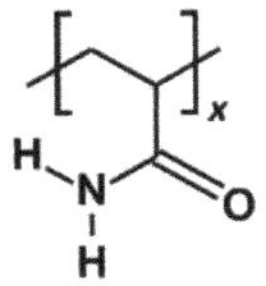	$x = 3$ (repeat unit consists of 3 acrylamide monomers)	213.91
HPAM 33%	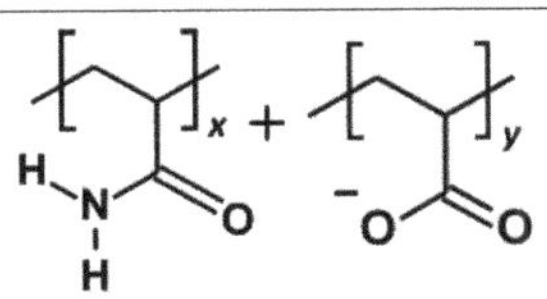	$x{:}y = 2{:}1$	213.89
SPAM 33%	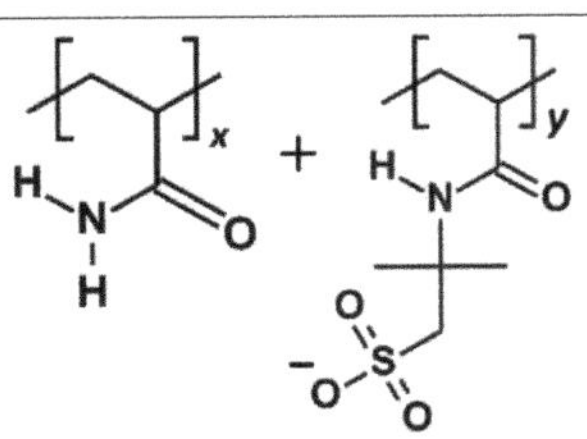	$x{:}y = 2{:}1$	372.06

3.4. System Preparation and Configuration

For the aqueous environment, 3000 water molecules were placed above the calcite surface, with a single polymer chain placed within the water solvent. HPAM 33%, the basis polymer model, consisted of 30 repeat units. To ensure electroneutrality in the simulation environment, 30 sodium ions were randomly placed within the solvent. A vacuum layer was added to the simulation box to isolate one of the calcite surfaces and direct the adsorption of the polymer onto a single surface layer. This setup prevents the competitive adsorption of the polymer to both surfaces during simulation, allowing for more consistent results analysis. The resulting simulation system had dimensions of a = 48.86 Å, b = 49.76 Å, and c = 86.38 Å, as illustrated in Figure 6.

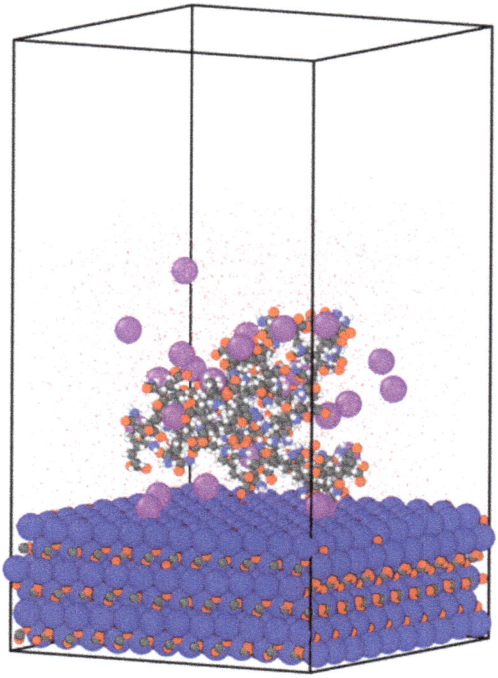

Figure 6. Initial configuration of the calcite–water system with HPAM 33% consisting of 30 repeat units. The sizes of the water molecules are reduced to tiny red dots for better visualisation. The remaining atoms are visualised with relative atom sizes and coloured based on element type. Calcite: Ca^{2+} (deep blue), carbon (grey), oxygen (red). HPAM: carbon (grey), oxygen (red), hydrogen (white), nitrogen (light blue). Na^{+} (purple).

Table 8 includes the designated parameter conditions and system descriptions for different parameters. In general, the simulation study for each parameter considered the lower, intermediate, and higher ranges to compare with experimental results. When varying a specific parameter, other parameters were kept constant at the control condition of HPAM 33% with 30 repeat units at 25 °C in a salt-free environment. All the experimental candidates were HPAM polymers, with labels representing different polymer chemistries, as detailed in the published work [55].

Table 8. Designated parameter spaces and system descriptions for adsorption studies with experimental conditions. Control system details: HPAM polymer with 30 repeat units and 33% charge density at 25 °C in salt-free conditions.

Parameter	System Description		MD Simulation	Experimental Condition
Polymer molecular weight (MW)	HPAM repeat units (213 Da/repeat unit)	Lower MW	10	F3330S (11–13 MDa)
		Middle MW	20	F3530S (15–17 MDa)
		Higher MW	30	F3630S (18–20 MDa)
Polymer charge density (CD)	HPAM charge density (%)	Lower CD	10	AN910 (10%)
		Middle CD	30	F3330S (30%)
		Higher CD	40	AN945 (40%)
Temperature (T)	System temperature (°C)	Lower T	25	F3330S, 25
		Middle T	60	F3330S, 80
		Higher T	100	F3330S, 95
Pressure (P)	System pressure (atm)	Lower P	1	-
		Middle P	80	-
		Higher P	160	-
Salinity	Saline environment with different cations but same anions, Cl^-	Monovalent cation	Li^+	-
		Monovalent cation	Na^+	F3530S, 3 wt% NaCl
		Monovalent cation	K^+	-
		Divalent cation	Mg^{2+}	-
		Divalent cation	Ca^{2+}	F3530S, 3 wt% $CaCl_2$
		Divalent cation	Sr^{2+}	-

When examining polymer chemistries such as polymer molecular weight and charge density, the HPAM structure was adjusted accordingly to mimic experimental candidates. For example, HPAM 33% with 10, 20, and 30 repeat units were constructed when investigating polymer molecular weight. In terms of polymer charge density, the polymer repeat unit was modified to match the charge density of the experimental candidates. HPAM chains with 10% (repeat unit consisting of nine acrylamide monomers and one acrylate monomer), 30% (repeat unit composed of seven acrylamide monomers and three acrylate monomers), and 40% (repeat unit consisting of six acrylamide monomers and four acrylate monomers) charge densities were constructed. Each chain had six repeat units, with the negative charges distributed evenly across the chains.

In some parameter conditions, such as high-pressure scenarios that are challenging to achieve experimentally, simulation results were analysed independently to provide predictive insights into the adsorption behaviours. Similarly, when assessing the effect of salinity, where experimental data are limited to NaCl and $CaCl_2$, simulations can explore the impact of salt ion valency with varying ion sizes on the adsorption mechanism. Three monovalent cations, lithium (Li^+), sodium (Na^+), and potassium (K^+), and three divalent cations, magnesium (Mg^{2+}), calcium (Ca^{2+}), strontium (Sr^{2+}), were modelled in this study, all paired with chloride anions (Cl^-). To recreate the three wt% NaCl solution in experiments, 29 NaCl molecules, equivalent to 0.53M, were placed in the solvent. The same number of cations and anions were placed for different cation cases to ensure consistency, with double the number of anions for divalent cations ($MgCl_2$, $CaCl_2$ and $SrCl_2$). Additionally, neutral PAM and SPAM 33% were investigated together with HPAM 33% in this parameter study, all with 30 repeat units, to observe the effect of polymer functional groups in a saline environment. The counterions placed in HPAM 33% and SPAM 33% systems consisted of the corresponding cations (30 and 15 counterions for monovalent and divalent cations, respectively).

3.5. Simulation Details and Analysis Methods

The system was equilibrated for each parameter using the NVT ensemble at 300 K for 500 ps, followed by a production run of 30 ns under the same simulation conditions. The only exception was during the investigation of pressure, where the simulation was performed under the NPT ensemble to adjust the pressure accordingly. The polymer adsorption behaviour was observed, and the adsorption amount on the calcite was computed. This adsorption amount, or adsorption surface density, was defined as the number of polymer chain atoms bound within a layer thickness of 4 Å above the calcite surface. The total number of adsorbed polymer atoms was then normalized by the calcite surface area.

We highlight that the simulation time during the production run was sufficient for polymer equilibration for qualitative analysis, as shown by the adsorption time profile in Figure 7 for the polymer charge density parameter. Only the averaged polymer adsorption amount was reported in the results section, averaged over the last 2 ns across five different realizations and compared with experimental results whenever available. As the adsorption simulations require extremely long times to reach true equilibrium, this relatively short sampling interval of 2 ns is chosen to minimize fluctuations and provides more consistent computed values. It is also noted that during the preliminary study, the results were not affected by the system size effect, as the adsorbed amount of polymer with 30 repeat units showed no significant change when placed into a larger calcite–water system. Furthermore, some water molecules could evaporate to the opposite surface under the vacuum slab during the simulation, forming a thin, stabilized water layer. Nevertheless, the number of water molecules evaporated to the surface slab was negligible and did not affect polymer adsorption density.

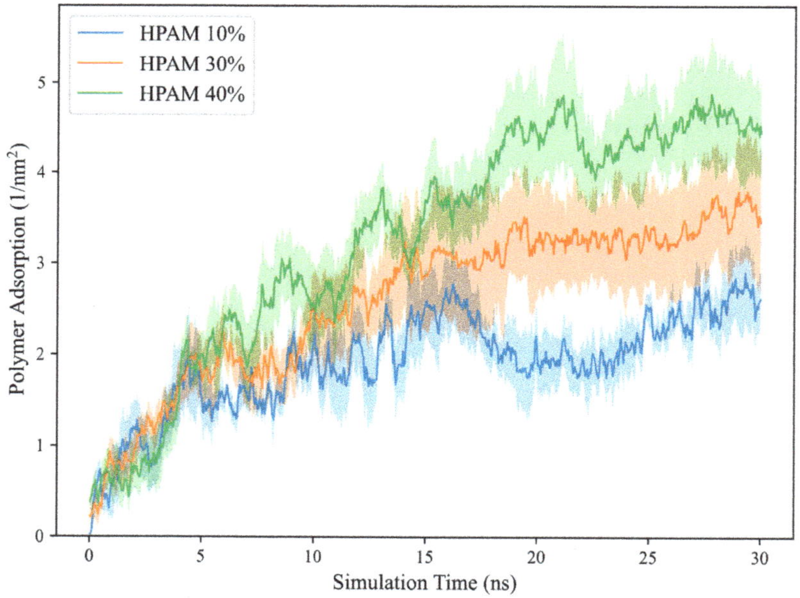

Figure 7. Polymer adsorption time profile for polymer charge density case. HPAM 10%, 30%, and 40% refer to polymers with charge densities of 10%, 30% and 40%, respectively. Shaded areas represent standard error from 5 realizations.

For the experimental methodologies, the polymer adsorption measurement workflow has been detailed in published works [6,39] and will not be reiterated here. The experimental adsorption data were taken from the equilibrium adsorbed value at the optimum polymer concentration [56]. Due to the differences in the scales and methodologies in quantifying polymer adsorption, it is important to note that the comparison between simulation and experimental results is purely qualitative with trend indications only.

The polymer conformation behaviour during the adsorption can be assessed across different parameters. Although the radius of gyration, R_g can be employed to measure the polymer size, it does not provide meaningful analysis of the polymer shape, such as whether they adopt an extended or coiled conformation. While the polymer shape may be interpreted as a flat, extended conformation with a larger R_g or a compact, coiled conformation with a smaller R_g under similar polymer structures, such analysis becomes difficult when assessing polymers of different sizes. A more useful property to characterise the shape is the relative shape anisotropy, κ^2. κ^2 can be computed from the gyration of tensor, S, adapted from [57]:

$$S = \frac{1}{N}\begin{pmatrix} \sum_i (x_i - x_C)^2 & \sum_i (x_i - x_C)(y_i - y_C) & \sum_i (x_i - x_C)(z_i - z_C) \\ \sum_i (y_i - y_C)(x_i - x_C) & \sum_i (y_i - y_C)^2 & \sum_i (y_i - y_C)(z_i - z_C) \\ \sum_i (z_i - z_C)(x_i - x_C) & \sum_i (z_i - z_C)(y_i - y_C) & \sum_i (z_i - z_C)^2 \end{pmatrix} \quad (7)$$

where subscript i represents atom i, C represents the centre of mass, and x, y, and z indicate the atomic coordinate in three directions, respectively. The eigenvalues of the matrix can be obtained by the diagonalization of S and are commonly sorted in descending order ($\lambda_1 > \lambda_2 > \lambda_3$).

$$S = diag(\lambda_1, \lambda_2, \lambda_3) \quad (8)$$

R_g and κ^2 can then be expressed as:

$$R_g^2 = \lambda_1 + \lambda_2 + \lambda_3 \quad (9)$$

$$\kappa^2 = 1 - 3\frac{(\lambda_1\lambda_2 + \lambda_2\lambda_3 + \lambda_3\lambda_1)}{(\lambda_1 + \lambda_2 + \lambda_3)^2} \quad (10)$$

κ^2 essentially characterises the dimensionality and symmetry of the polymer, ranging from 0 to 1. A value of 0 signifies a perfectly spherical shape, indicating a more coiled polymer conformation, while a value of 1 denotes a linear chain, reflecting a more extended conformation. It is noted that these terms are typically employed to describe a free polymer in solution. During polymer adsorption, the surface restricts the polymer's movement and distribution, effectively reducing its conformation freedom and causing it to exist in quasi-two-dimensional space. In the current study, the adsorbed polymer conformation may still be described with Equations (9) and (10), but with a slightly different interpretation.

4. Conclusions

This study has employed molecular dynamics simulations to investigate the adsorption behaviour of hydrolysed polyacrylamide (HPAM) onto calcite surfaces under various reservoir conditions. Our findings demonstrate that HPAM adsorption is enhanced by higher molecular weight, charge density, and temperature, while pressure exerts minimal influence. The interplay between salt-bridging and charge-screening mechanisms governs the impact of salinity on adsorption, with salt bridging being more pronounced for HPAM. While the simulation results generally correlate with experimental observations, discrepancies may arise from limitations in the simulation model and forcefield [58]. Overall, these findings provide valuable insights into the design and optimisation of polymer-based solutions for solids production control in carbonate reservoirs.

Supplementary Materials: The following supporting information can be downloaded at: https://www.mdpi.com/article/10.3390/molecules30020285/s1, S1. Forcefield Structure and Parameters Info, S2. Electrostatic Surface Potential Diagram for Possible Polymer-Calcite Adsorption Mechanism, S3. Density Profile Distribution of Cation and Polymer Atoms During Adsorption on Calcite Surface.

Author Contributions: Conceptualization, O.K.M., P.F.L. and E.A.M.; formal analysis, P.F.L. and E.A.M.; investigation, K.Y.H. and J.H.L.; methodology, K.Y.H.; project administration, O.K.M.; supervision, P.F.L. and E.A.M.; writing—original draft, K.Y.H.; writing—review and editing, O.K.M., P.F.L. and E.A.M. All authors have read and agreed to the published version of the manuscript.

Funding: This research was funded by PETRONAS (Malaysia) through the PETRONAS Centre for Engineering of Multiphase Systems.

Data Availability Statement: Data supporting this manuscript, including example input files to perform equilibrium adsorption across different parameters can be downloaded at Figshare (DOI://10.6084/m9.figshare.27890982).

Acknowledgments: We gratefully acknowledge Myo Thant Maung Maung, employed by PETRONAS for acquisition of funding and general supervision of this research.

Conflicts of Interest: The research was conducted without any commercial or financial relationships that could be construed as a potential conflict of interest. The funders had no role in the design of the study, in the collection, analyses, or interpretation of data, in the writing of the manuscript, or in the decision to publish the results.

References

1. Mowar, S.; Zaman, M.; Stearns, D.W.; Roegiers, J.C. Micro-mechanisms of pore collapse in limestone. *J. Pet. Sci. Eng.* **1996**, *15*, 221–235. [CrossRef]
2. Soares, A.C.; Altoé, J.E.; Bedrikovetsky, P.; Ferreira, F.H. Formation Damage due to Pore Collapse During Pressure Depletion. In Proceedings of the SPE European Formation Damage Conference, The Hague, The Netherlands, 13–14 May 2003.
3. Zaman, M.; Roegiers, J.C.; Abdulraheem, A.; Azeemuddin, M. Pore Collapse in Weakly Cemented and Porous Rocks. *J. Energy Resour. Technol.* **1994**, *116*, 97–103. [CrossRef]
4. Al-Awad, M.N.J.; El-Sayed, A.-A.H.; Desouky, S.E.-D.M. Factors Affecting Sand Production from Unconsolidated Sandstone Saudi Oil and Gas Reservoir. *J. King Saud. Univ.-Eng. Sci.* **1999**, *11*, 151–172. [CrossRef]
5. Alakbari, F.S.; Mohyaldinn, M.E.; Muhsan, A.S.; Hasan, N.; Ganat, T. Chemical Sand Consolidation: From Polymers to Nanoparticles. *Polymers* **2020**, *12*, 1069. [CrossRef] [PubMed]
6. Hue, K.Y.; Lew, J.H.; Myo Thant, M.M.; Matar, O.K.; Luckham, P.F.; Müller, E.A. Molecular Dynamics Simulation of Polyacrylamide Adsorption on Calcite. *Molecules* **2023**, *28*, 6367. [CrossRef] [PubMed]
7. Peng, P.; Garnier, G. Effect of cationic polyacrylamide on precipitated calcium carbonate flocculation: Kinetics, charge density and ionic strength. *Colloids Surf. A Physicochem. Eng. Asp.* **2012**, *408*, 32–39. [CrossRef]
8. Rasteiro, M.G.; Pinheiro, I.; Ahmadloo, H.; Hunkeler, D.; Garcia, F.A.P.; Ferreira, P.; Wandrey, C. Correlation between flocculation and adsorption of cationic polyacrylamides on precipitated calcium carbonate. *Chem. Eng. Res. Des.* **2015**, *95*, 298–306. [CrossRef]
9. Lee, B.; Schlautman, M. Effects of Polymer Molecular Weight on Adsorption and Flocculation in Aqueous Kaolinite Suspensions Dosed with Nonionic Polyacrylamides. *Water* **2015**, *7*, 5896–5909. [CrossRef]
10. Ali, M.; Mahmud, H.B. The effects of Concentration and Salinity on Polymer Adsorption Isotherm at Sandstone Rock Surface. *IOP Conf. Ser. Mater. Sci. Eng.* **2015**, *78*, 012038. [CrossRef]
11. Abdel-Azeim, S.; Kanj, M.Y. Dynamics, Aggregation, and Interfacial Properties of the Partially Hydrolyzed Polyacrylamide Polymer for Enhanced Oil Recovery Applications: Insights from Molecular Dynamics Simulations. *Energy Fuels* **2018**, *32*, 3335–3343. [CrossRef]
12. Mintis, D.G.; Mavrantzas, V.G. Effect of pH and Molecular Length on the Structure and Dynamics of Short Poly(acrylic acid) in Dilute Solution: Detailed Molecular Dynamics Study. *J. Phys. Chem. B* **2019**, *123*, 4204–4219. [CrossRef] [PubMed]
13. Yang, X.; Xiong, Z.; Mao, J.; Yang, T.; Fu, B.; Han, D.; Yang, J.; Chen, W.; Liu, W.; Zhang, Q.; et al. Molecular dynamics simulations to study the adsorption damage of modified polyacrylamide in sandstone pores. *J. Mol. Liq.* **2024**, *397*, 124096. [CrossRef]
14. Carrillo, J.-M.Y.; Dobrynin, A.V. Molecular Dynamics Simulations of Polyelectrolyte Adsorption. *Langmuir* **2007**, *23*, 2472–2482. [CrossRef]
15. Willemsen, J.A.R.; Myneni, S.C.B.; Bourg, I.C. Molecular Dynamics Simulations of the Adsorption of Phthalate Esters on Smectite Clay Surfaces. *J. Phys. Chem. C* **2019**, *123*, 13624–13636. [CrossRef]

16. Ma, X.; Sun, X.; Chang, M.; Liu, Q.; Dong, X.; Fan, Y.; Chen, R. Adsorption of Different Ionic Types of Polyacrylamide on Montmorillonite Surface: Insight from QCM-D and Molecular Dynamic Simulation. *Molecules* **2023**, *28*, 4417. [CrossRef] [PubMed]
17. Qu, Y.; Lin, L.; Gao, S.; Yang, Y.; Huang, H.; Li, X.; Ren, H.; Luo, W. A molecular dynamics study on adsorption mechanisms of polar, cationic, and anionic polymers on montmorillonite. *RSC Adv.* **2023**, *13*, 2010–2023. [CrossRef] [PubMed]
18. Quezada, G.R.; Leiva, W.; Saavedra, J.H.; Robles, P.; Galvez, E.; Jeldres, R.I. A Molecular Dynamics Simulation of Polymers' Interactions with Kaolinite (010) Surfaces in Saline Solutions. *Polymers* **2022**, *14*, 3851. [CrossRef] [PubMed]
19. Choi, I.; Kim, I.W. Molecular Dynamics Simulation to Understand the Ability of Anionic Polymers to Alter the Morphology of Calcite. *Int. J. Polym. Sci.* **2017**, *2017*, 5. [CrossRef]
20. Huang, W.; Geng, X.; Li, J.; Zhou, C.; Liu, Z. Molecular Dynamics Study on the Adsorption and Modification Mechanism of Polymeric Sand-Fixing Agent. *Polymers* **2022**, *14*, 3365. [CrossRef] [PubMed]
21. Ghatee, M.H.; Koleini, M.M.; Ayatollahi, S. Molecular dynamics simulation investigation of hexanoic acid adsorption onto calcite $(10\bar{1}4)$surface. *Fluid. Phase Equilibria* **2015**, *387*, 24–31. [CrossRef]
22. Krishna, S.; Sreedhar, I.; Patel, C.M. Molecular dynamics simulation of polyamide-based materials—A review. *Comput. Mater. Sci.* **2021**, *200*, 110853. [CrossRef]
23. Quezada, G.R.; Rozas, R.E.; Toledo, P.G. Polyacrylamide adsorption on (1 0 1) quartz surfaces in saltwater for a range of pH values by molecular dynamics simulations. *Miner. Eng.* **2021**, *162*, 106741. [CrossRef]
24. Sparks, D.J.; Romero-Gonzalez, M.E.; El-Taboni, E.; Freeman, C.L.; Hall, S.A.; Kakonyi, G.; Swanson, L.; Banwart, S.A.; Harding, J.H. Adsorption of poly acrylic acid onto the surface of calcite: An experimental and simulation study. *Phys. Chem. Chem. Phys.* **2015**, *17*, 27357–27365. [CrossRef] [PubMed]
25. Wang, S.; Li, G.; Li, Y.; Guo, J.; Zhou, S.; Yong, S.; Pan, B.; Bai, B. Adsorption of new hydrophobic polyacrylamide on the calcite surface. *J. Appl. Polym. Sci.* **2017**, *134*, 45314. [CrossRef]
26. Ji, Y.X.; Wang, F.H.; Duan, L.C.; Zhang, F.; Gong, X.D. Effect of temperature on the adsorption of sulfanilamide onto aluminum oxide and its molecular dynamics simulations. *Appl. Surf. Sci.* **2013**, *285*, 403–408. [CrossRef]
27. Wisniewska, M.; Chibowski, S.; Urban, T.; Sternik, D.; Terpilowski, K. Impact of anionic polyacrylamide on stability and surface properties of the Al2O3-polymer solution system at different temperatures. *Colloid. Polym. Sci.* **2016**, *294*, 1511–1517. [CrossRef] [PubMed]
28. Ahsani, T.; Tamsilian, Y.; Rezaei, A. Molecular dynamic simulation and experimental study of wettability alteration by hydrolyzed polyacrylamide for enhanced oil recovery: A new finding for polymer flooding process. *J. Pet. Sci. Eng.* **2021**, *196*, 108029. [CrossRef]
29. Chen, H.; Panagiotopoulos, A.Z.; Giannelis, E.P. Atomistic molecular dynamics simulations of carbohydrate-calcite interactions in concentrated brine. *Langmuir* **2015**, *31*, 2407–2413. [CrossRef]
30. Quezada, G.R.; Jeldres, R.I.; Fawell, P.D.; Toledo, P.G. Use of molecular dynamics to study the conformation of an anionic polyelectrolyte in saline medium and its adsorption on a quartz surface. *Miner. Eng.* **2018**, *129*, 102–105. [CrossRef]
31. Quezada, G.R.; Toro, N.; Saavedra, J.; Robles, P.; Salazar, I.; Navarra, A.; Jeldres, R.I. Molecular Dynamics Study of the Conformation, Ion Adsorption, Diffusion, and Water Structure of Soluble Polymers in Saline Solutions. *Polymers* **2021**, *13*, 3550. [CrossRef]
32. Carnal, F.; Stoll, S. Adsorption of weak polyelectrolytes on charged nanoparticles. Impact of salt valency, pH, and nanoparticle charge density. Monte Carlo simulations. *J. Phys. Chem. B* **2011**, *115*, 12007–12018. [CrossRef] [PubMed]
33. Chen, P.; Yao, L.; Liu, Y.; Luo, J.; Zhou, G.; Jiang, B. Experimental and theoretical study of dilute polyacrylamide solutions: Effect of salt concentration. *J. Mol. Model.* **2012**, *18*, 3153–3160. [CrossRef] [PubMed]
34. Wang, S.; Zhang, L.; Yan, B.; Xu, H.; Liu, Q.; Zeng, H. Molecular and Surface Interactions between Polymer Flocculant Chitosan-g-polyacrylamide and Kaolinite Particles: Impact of Salinity. *J. Phys. Chem. C* **2015**, *119*, 7327–7339. [CrossRef]
35. Sun, W.; Zeng, H.; Tang, T. Enhanced Adsorption of Anionic Polymer on Montmorillonite by Divalent Cations and the Effect of Salinity. *J. Phys. Chem. A* **2021**, *125*, 1025–1035. [CrossRef] [PubMed]
36. Yan, C.; Kramer, P.L.; Yuan, R.; Fayer, M.D. Water Dynamics in Polyacrylamide Hydrogels. *J. Am. Chem. Soc.* **2018**, *140*, 9466–9477. [CrossRef] [PubMed]
37. Dai, S.; Liu, Y.; Huang, Z.; Zhao, X. Molecular dynamics simulations on the interaction between microsphere and water in nanosilica/crosslinked polyacrylamide microsphere aqueous solution with a core–shell structure and its swelling behavior. *Compos. Interfaces* **2017**, *25*, 69–92. [CrossRef]
38. Ni, T.; Huang, G.-S.; Zheng, J.; Gao, P.; Chen, M.-M. Research on the crosslinking mechanism of polyacrylamide/resol using molecular simulation and X-ray photoelectron spectroscopy. *Polym. J.* **2010**, *42*, 357–362. [CrossRef]
39. Lew, J.H.; Matar, O.K.; Müller, E.A.; Maung, M.T.M.; Luckham, P.F. Adsorption of Hydrolysed Polyacrylamide onto Calcium Carbonate. *Polymers* **2022**, *14*, 405. [CrossRef] [PubMed]
40. Bolto, B.; Gregory, J. Organic polyelectrolytes in water treatment. *Water Res.* **2007**, *41*, 2301–2324. [CrossRef] [PubMed]

41. Tanner, J.J. Empirical power laws for the radii of gyration of protein oligomers. *Acta Crystallogr. D Struct. Biol.* **2016**, *72*, 1119–1129. [CrossRef]
42. Zhou, Y.; Yu, H.; Wanless, E.J.; Jameson, G.J.; Franks, G.V. Influence of polymer charge on the shear yield stress of silica aggregated with adsorbed cationic polymers. *J. Colloid. Interface Sci.* **2009**, *336*, 533–543. [CrossRef]
43. Hamal, P.; Subasinghege Don, V.; Nguyenhuu, H.; Ranasinghe, J.C.; Nauman, J.A.; McCarley, R.L.; Kumar, R.; Haber, L.H. Influence of Temperature on Molecular Adsorption and Transport at Liposome Surfaces Studied by Molecular Dynamics Simulations and Second Harmonic Generation Spectroscopy. *J. Phys. Chem. B* **2021**, *125*, 10506–10513. [CrossRef] [PubMed]
44. Humbert, M.T.; Zhang, Y.; Maginn, E.J. PyLAT: Python LAMMPS Analysis Tools. *J. Chem. Inf. Model.* **2019**, *59*, 1301–1305. [CrossRef] [PubMed]
45. Quezada, G.; Rozas, R.; Toledo, P. Molecular dynamics simulations of the adsorption of partially hydrolyzed polyacrylamide on kaolinite edge surfaces in saltwater: Effect of temperature. In Proceedings of the 12th European Congress of Chemical Engineering & 5th European Congress of Applied Biotechnology, Florence, Italy, 15–19 September 2019.
46. Briscoe, B.; Luckham, P.; Zhu, S. Pressure influences upon shear thickening of poly(acrylamide) solutions. *Rheol. Acta* **1999**, *38*, 224–234. [CrossRef]
47. Bai, S.; Kubelka, J.; Piri, M. A positively charged calcite surface model for molecular dynamics studies of wettability alteration. *J. Colloid. Interface Sci.* **2020**, *569*, 128–139. [CrossRef]
48. Okur, H.I.; Hladilkova, J.; Rembert, K.B.; Cho, Y.; Heyda, J.; Dzubiella, J.; Cremer, P.S.; Jungwirth, P. Beyond the Hofmeister Series: Ion-Specific Effects on Proteins and Their Biological Functions. *J. Phys. Chem. B* **2017**, *121*, 1997–2014. [CrossRef]
49. Ali, A.; Le, T.T.B.; Striolo, A.; Cole, D.R. Salt Effects on the Structure and Dynamics of Interfacial Water on Calcite Probed by Equilibrium Molecular Dynamics Simulations. *J. Phys. Chem. C* **2020**, *124*, 24822–24836. [CrossRef]
50. Lew, J.H.; Hue, K.Y.; Matar, O.K.; Müller, E.A.; Luckham, P.F.; Santos, A.S.; Myo Thant, M.M. Atomic Force Microscopy and Molecular Dynamic Simulation of Adsorption of Polyacrylamide with Different Chemistries onto Calcium Carbonate. *Polymers* **2024**, *16*, 494. [CrossRef]
51. Materials, D. MedeA 3.5 (Materials Exploration and Design Analysis). 2022. Available online: https://www.materialsdesign.com/medea-software (accessed on 21 November 2024).
52. Plimpton, S. Fast Parallel Algorithms for Short-Range Molecular Dynamics. *J. Comput. Phys.* **1995**, *117*, 1–19. [CrossRef]
53. Hockney, R.W.; Eastwood, J.W. *Computer Simulation Using Particles*; Adam Hilger: New York, NY, USA, 1989.
54. Martyna, G.J.; Tobias, D.J.; Klein, M.L. Constant pressure molecular dynamics algorithms. *J. Chem. Phys.* **1994**, *101*, 4177–4189. [CrossRef]
55. Lew, J.H.; Luckham, P.F.; Matar, O.K.; Müller, E.A.; Santos, A.S.; Maung Maung, M.T. Consolidation of Calcium Carbonate Using Polyacrylamides with Different Chemistries. *Powders* **2023**, *3*, 1–16. [CrossRef]
56. Lew, J.H. An experimental study of calcium carbonate consolidation via polyacrylamide for carbonate reservoir strengthening. PhD. Thesis, Imperial College London, London, UK, 2024.
57. Headen, T.F.; Boek, E.S.; Jackson, G.; Totton, T.S.; Müller, E.A. Simulation of Asphaltene Aggregation through Molecular Dynamics: Insights and Limitations. *Energy Fuels* **2017**, *31*, 1108–1125. [CrossRef]
58. Tillotson, M.J.; Diamantonis, N.I.; Buda, C.; Bolton, L.W.; Müller, E.A. Molecular modelling of the thermophysical properties of fluids: Expectations, limitations, gaps and opportunities. *Phys. Chem. Chem. Phys.* **2023**, *25*, 12607–12628. [CrossRef] [PubMed]

molecules

MDPI

Article

Simulation of Solvatochromic Phenomena in Xanthione Using Explicit Solvent Methods

Anjay Manian [1,*], Zifei Chen [1,2], Rohan J. Hudson [2,3] and Salvy P. Russo [1,*]

1 ARC Centre of Excellence in Exciton Science, School of Science, RMIT University, Melbourne 3000, Australia; zifei.chen@rmit.edu.au
2 ARC Centre of Excellence in Exciton Science, School of Chemistry, University of Melbourne, Parkville 3010, Australia; rohan.hudson@moglabs.com
3 MOGLabs, Carlton 3053, Australia
* Correspondence: anjay.manian3@rmit.edu.au (A.M.); salvy.russo@rmit.edu.au (S.P.R.)

Abstract: Xanthione is a sulfated polycyclic aromatic hydrocarbon which exhibits unique anti-Kasha properties and substantial sensitivity to its medium. Due to this sensitivity however, this makes xanthione-based systems very difficult to simulate. Further, xanthione's is understood to be come more photostable in the presence of a highly polar medium, however whether these photophysical properties could be taken advantage of for certain applications remains to be seen. In clarifying long-held beliefs of specific solvent effects, we apply a rigorous theoretical solvent analysis in both implicit and explicit solvent mediums to elucidate a more complete description of solvent polarity sensitivity in xanthione using both quantum chemical and molecular dynamics techniques. Not only was it found that explicit solvation methods are vital in an accurate description of the system, only a handful of explicit solvent molecules in the simulation are required to yield an appropriate electronic description. This short work is vital to devising future applications for xanthione-based and other quantum technologies, and is an important foundation stone on this journey.

Keywords: xanthione; solvent effects; explicit solvation; DFT; MRCI; MD

1. Introduction

Xanthones are polycyclic aromatic hydrocarbon which have been shown to display tuneable photophysical properties with respect to their atomic surroundings [1–3]. Indeed, a shift in photoluminescence quantum yield from 0.0002 in cyclohexane to 0.46 in trifluoroethanol is observable [4], highlighting a distinct solvatochromic nature. The 1980s held the most interest for sulfated xanthones, or xanthiones, however interest eventually waned after it was concluded that there were some solvent-specific effects influencing excited-state properties. However, more recently interest has begun to increase again due to their anti-Kasha optical properties and suspected singlet fission qualities [5], and is also solvatochromic in nature [6,7], with new studies being reported regularly. However, experimental testing is both time consuming and financially expensive; therefore a theoretical treatment may be more prudent.

However, accurate theoretical modelling of solvatochromic systems is often difficult if not impossible; due primarily to implicit solvent models like the polarisable continuum model (PCM), whereby the solvent is approximated using a dielectric cavity, not capturing the important electron structure properties that comes with solute-solvent bonding interactions. This phenomenon is observable in a myriad of systems, such as in indole [8,9], 2.4-bismidazolylphenol [10], and naphthalene diimide [11,12].

There is also the question of how complete the explicit solvation need be. Our previous work on indole [8] highlighted that there is typically some saturation limit on the number of solvent molecules that need to be included in the simulation. If we take a given xanthione chromophore, it is not clear how many solvent molecules are required to replicate experimental conditions. In addition, singlet fission is highly sensitive to the energy of the low-lying excited states [13]; if we consider that both the optically dark and bright states can be influenced by the polarity of a given solvent, can we use this information to fuel future xanthione-singlet fission studies?

This work will seek to gain further clarity concerning these important questions. This short but vital work will use 3,6-bis(diethylamino)-xanthene-9-thione as our xanthione derivative (Figure 1) to elucidate how solvent polarity effects the energies of both the optically dark and bright states, referred hereafter to as the L_a and L_b state, respectively (similar to Platt's notation [14]). Vertical excitation energies at both Franck-Condon points as well as the electronic structure will be calculated within implicit and explicit solvent models, in order to further understand how solvent polarity alters the photophysical properties of xanthione.

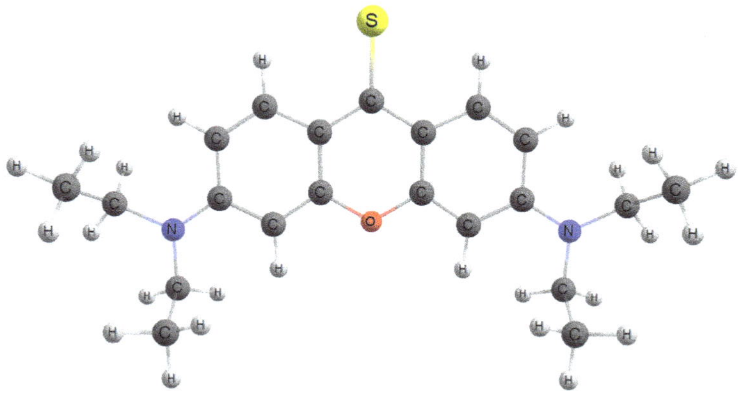

Figure 1. The xanthione compound used in this work. Its detailed name is 3,6-bis(diethylamino)-xanthene-9-thione.

2. Computational Details

Geometries for the electronic ground S_0 state, and the first and second electronic excited L_a and L_b states, were all calculated using the Coulomb-attenuated Becke 3-parameter Lee–Yang–Parr (CAM-B3LYP) [15–19] exchange-correlation functional with the damped Becke-Johnson geometry dependent 3-parameter DFT-D3(BJ) dispersion correction [20,21], alongside the double-ζ 6-primitive Slater-type orbital 3-inner and 1-outer 6–31G basis set [22–24], as implemented in the GAUSSIAN 16 software package [25]. Implicit solvent systems (xanthione only) were simulated using a PCM via a ground state approach. Here, the self-consistent field is corrected through a solvent-effects term, allowing for optimisation within solution of excited states. Explicit solvent models were optimised by placing the xanthione alongside explicit solvent molecules, in addition to a PCM. Explicit solvation was performed in sets of 2, 4, and 6, i.e., xanthione and n-solvent molecules within a PCM.

Initial geometries were chosen based on a simplified charge analysis. Following optimisation of the gas phase geometries, 2 solvent molecules were injected into the system above and below the xanthione plane favouring the sulfur-site. Following successful optimisation, 2 more were injected above and below the oxygen-site, and following optimisation 2 more on either side near the nitrogen-centres. This was done for each solute-solvent system.

To investigate the effect of solvent polarity on the excited-state manifold of xanthione, solvents were selected primarily based on their dielectric constant, rather than their solubility with respect to xanthione. While solubility is typically a key factor in solvent selection for such studies, the focus on dielectric constant allowed for a broader exploration of solvent polarity effects. It should be noted, however, that some of the solvents used in this work, though less commonly employed with xanthione due to solubility limitations, still provide valuable insights into the solute-solvent interaction of xanthione. The solvents used in this work are listed in Table 1.

Table 1. Solvents, and their respective dielectric constants, paired with xanthione used in this work.

Solvent	Dielectric Constant
Toluene	2.374
1,2-Dichloroethane	10.125
Acetone	20.493
Acetonitrile	35.688
Dimethyl Sulfoxide	46.826
Formic acid	51.100
Water	78.355
Formamide	108.940

All dielectric constants reported in this work are taken as per GAUSSIAN16 documentation [25].

Single-point calculations for each optimised structure was performed using the DFT-based multireference configuration interaction DFT/MRCI method [26,27]. The one-particle basis was calculated using the the Becke "half-and-half" Lee-Yang-Par BHLYP functional [28,29] with the DFT-D3(BJ) correction and the Karlsruhe variant of the split valence with polarisation functions on non-hydrogen atoms def2-SV(P) basis set [30,31], as implemented in the TURBOMOLE software package [32]. The DFT/MRCI reference space was generated iteratively by including all electron configurations with expansion coefficients greater than 10^{-3} in numerous probe calculations, using 10 electrons across 10 orbitals, allowing only for a maximum of two-electron excitations. Probe runs were calculated by discarding configurations with energy less than the highest reference energy; starting with a threshold of 0.6 E_h, then 0.8, with the finalised wavefunction built using a threshold of 1.0. Molecular orbitals with energies larger than 2.0 E_h were not used. We note that n = 4, 6 in toluene was not resolved due to computational limitations.

To further demonstrate the impact the chosen number of solvent molecules has in explicit solvent models, we also perform molecular dynamics (MD) simulations. Here, we have chosen four typical solvents: toluene, acetone, dimethyl sulfoxide (DMSO), and water. The simulation system included one xanthione molecule and 400 solvent molecules (800 in the case of water). Periodic boundary conditions are employed. First, a 5 ns isobaric-isothermal (NPT) ensemble simulation was conducted to equilibrate the system at room temperature, yielding stable boxes with lengths 42.5 Å, 37.6 Å, 37.2 Å, and 29.2 Å, for each of the 4 solvents respectively. This was followed by a 5 ns canonical (NVT) ensemble simulation to sample the system for statistical analysis. The MD simulations were carried out using the GROMACS software package [33], employing the General Amber Force Field (GAFF) [34] for both solute and solvent. Effective charges were calculated using the restrained electrostatic potential (RESP) fitting method [35,36], using the B3LYP density functional and the Karlsruhe variant of the triple-ζ valence polarised def2-TZVP basis set [31,37]. Electrostatic interactions were treated using the particle-mesh Ewald (PME) method [38], while Lennard-Jones interactions were cut off at 8 Å. Temperature control was maintained via the Nosé–Hoover thermostat [39,40], and pressure was regulated in the NPT simulations using the Berendsen barostat [41].

3. Results & Discussion

For xanthione, a dominant leading configuration state function (CSF) of one below the highest occupied molecular orbital (HOMO-1) to the lowest unoccupied molecular orbital (LUMO) is observed for the L_a state, while for the L_b state a dominant HOMO→LUMO CSF is noted. In the gas phase, there is some level inversion between the non-bonding $n\pi^\star$ state and a nearby $\sigma\pi^\star$ state, however upon solvation (independent of polarity) the L_a state is of $n\pi^\star$ character. In all cases, the L_b state is bright, and is of $\pi\pi^\star$ character. These state assignments agree well with reported chemistry [42] in 3-methylpentane (ϵ = 1.895 [43]).

Examination of the fluorescence qualities of xanthione in various solvents shows that for the most part PCMs tend to overestimate vertical excitation energies with respect to the trends observed from explicit solvation. The absorption energies (Figure 2) of the L_b state show implicit solvent energies to trend upwards from around 2.35 eV and plateaus at ∼2.57 eV. X = 2 explicit solvation trends lower in energy by more than 0.1 eV, with larger solvent shells of similar quality, but progressively lower. Outliars to this trend are observable for X = 2, 4 in dichloroethane, X = 2 in acetone, and X = 6 in dimethyl sulfoxide. Like the L_b states, PCM energies are overestimated by 0.8 eV for the L_a manifold, but no large outliars are observable in the data, with a relatively smooth progression in energies. The emission profiles are significantly more variable compared to the absorption spectra; while implicit solvents models yield overestimated energies as with the absorption spectra, the difference is larger: around 0.15 eV on the L_b manifold, and up to 0.6 eV on the L_a manifold. Interesting, most of the data points in the lower solvent polarity region appear to not follow any clear trend. We suspect that this is due to solvent drift (since the solvent is not frozen between optimisations, small changes may have large impacts on the electronic structure, and therefore the energies), however of note is a large disagreement for X = 4, 6 in dimethyl sulfoxide on both L_a and L_b manifolds.

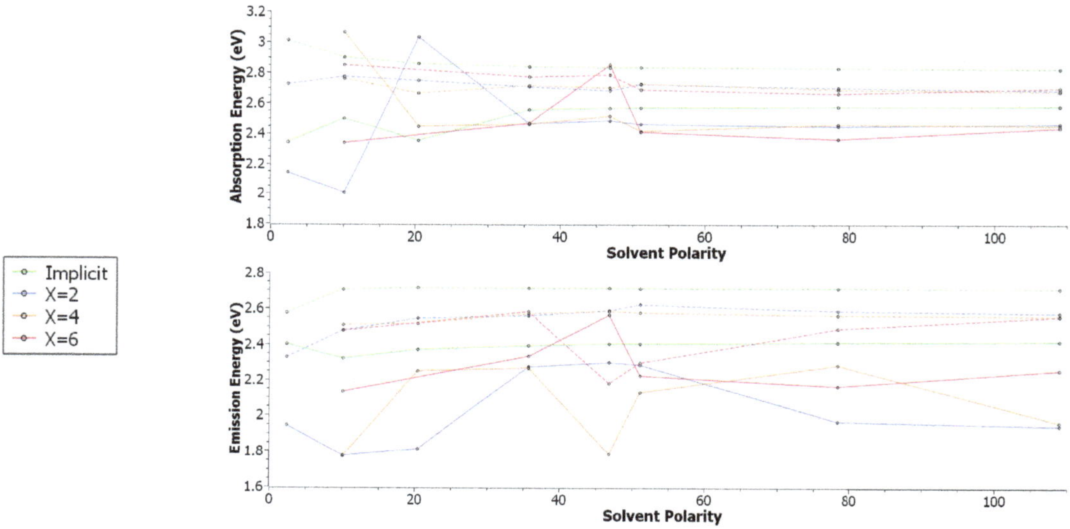

Figure 2. Xanthione vertical absorption and emission energies (DFT/MRCI) calculated for a polarisable continuum model and with X number of explicit solvent molecules in addition to a polarisable continuum model, as a function of the solvent polarity. Solid lines refer to the L_a state, while dashed lines refer to the L_b state.

It is very important to note that for many of these systems, direct absorption to the L_a manifold is more energetically favourable than excitation to the L_b state. in other words, there is no adiabatic level inversion between the L_a and L_b state in xanthione noted in this work (Figure 3), however there is a large reorganisation energy in some of these compounds, likely due again to solvent drift.

A closer inspection of the electronic structure (see ESI) shows that while implicit solvents show identical electron densities for both L_a and L_b states, explicit solvation results in some exciton delocalisation effects through solute-solvent interactions for the L_a state specifically. Indeed, for X = 2 in toluene for example, a small portion of the density can be observed shunted across to the solvent molecules. This is slightly more pronounced in X = 2 in acetone, but is most prevalent in X = 6 dimethyl sulfoxide. This behaviour does not appear in either the $n\pi^{\star}$ or $\pi\pi^{\star}$ manifolds. If we note the systems studied in this work as snapshots of an aggregate system studied in experiment, we can begin to understand why some data points in Figure 2 appear erroneous; these geometries can be considered as lower or upper bounds of an aggregate energetic profile.

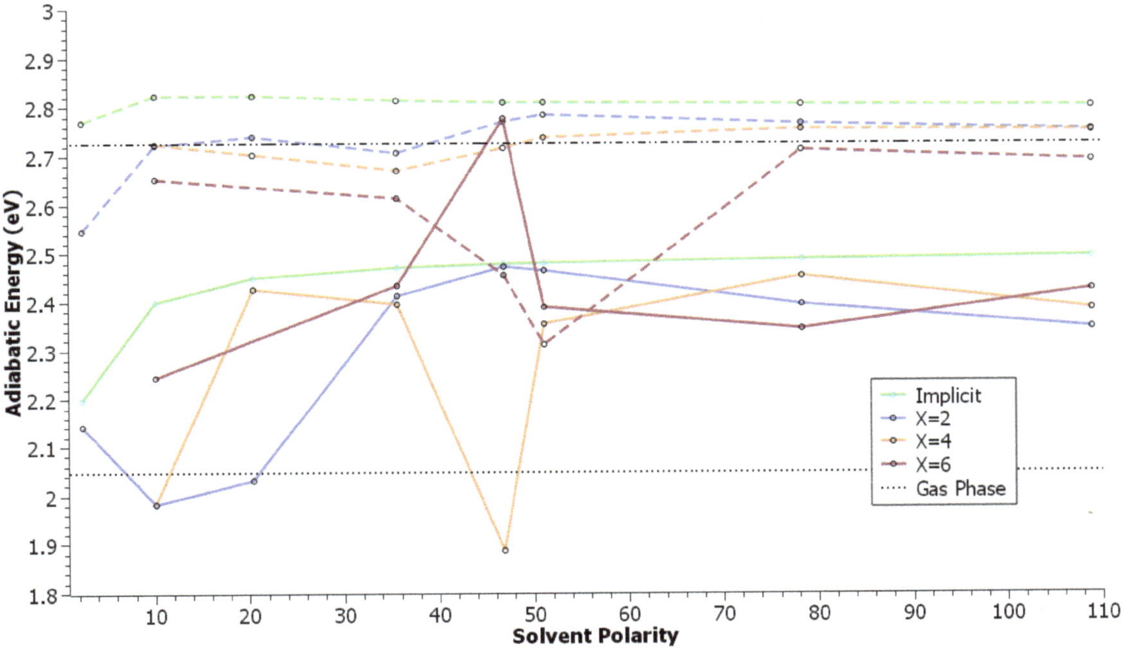

Figure 3. Calculated adiabatic energies (DFT/MRCI) for Xanthione in a polarisable continuum model and with X number of explicit solvent molecules in addition to a polarisable continuum model, as a function of the solvent polarity. Solid lines refer to the L_a state, while dashed lines refer to the L_b state. Black lines show gas phase energies for the L_a (dashed) and L_b (dot-dashed) states.

Despite these complications, these results correlate fairly well with experiment. Assuming L_b emission, Bondarev and co-workers [2] for example, for unsubstituted xanthione in n-hexane (ϵ = 1.882) report absorption and emission energies of 3.08 eV and 2.76 eV, respectively (2.73 eV and 2.33 eV in this work for toluene), and in acetonitrile energies of 3.05 eV and 2.73 eV, respectively (2.71 eV and 2.56 eV in this work). Lorenc and co-workers [44] reported similar energies in acetonitrile, with absorption and emission energies of 3.02 eV and 2.73 eV, respectively. Maciejewski & Steer [45] report the absorption energy to be between 2.97–3.65 eV in perfluoro-n-hexane (ϵ = 1.76 [46]) depending on the derivative, and a $S_2 \rightarrow S_1$ energy gap between 0.99–1.36 eV (0.40 ev in this work for toluene). This agrees with Capitanio and co-workers [42] who reported an absorption energy of 2.93 eV in 3-methylpentane (ϵ = 1.90 (from the Hazardous Substances Data Bank)). Maciejewski and co-workers [47] also report on absorption and emission characteristics in a myriad of different mediums; absorption energies of 3.16 eV in perfluoro-n-hexane, 2.91 eV in water (2.70 eV in this work), and 2.98 eV in micellar sodium perfluorooctanoate, and emission energies of 2.76 eV in perfluoro-n-hexane, and 2.67 eV in micellar sodium perfluorooctanoate (we are unable to source a dielectric constant for this medium, however as it is a micellar system we suspect aggregation to play an important role). There is one notable disagreement with literature, related to the $n\pi^*$ state: Capitanio and co-workers [42] also report a 3-methylpentane L_b absorption energy of 1.96 eV, which is 0.3 eV lower than any trends observed in this work. We suspect this is due to aggregation-induced energy level splitting of xanthione in solution.

From these results, we can see that concerning how much computational effort is required for explicit solvent methods, X = 2, 4 is typically enough to yield a good return on accuracy with respect to computational cost; X = 6 is just not necessary as the quality in results between X = 4 compared to X = 6 is within 10% of each other. Importantly, one needs to question whether the additional computational cost is worth that additional resolution. This data suggests that energies are most strongly influenced by how the solvent interacts with the xanthione core. Specifically, in the case of the highly polar solvents, hydrogen bonding can be observed to be commonplace. Accordingly, this behaviour is not captured by implicit models, but is by explicit models of X = 2, 4. However and as already mentioned, the configurations studied in this work as snapshots of a dynamic aggregate, and may not be representative of the molecular ensemble. In terms of some ensemble average, it is difficult to know for certain how the configurations studied in this work relate to this ensemble average, and is a very common problem when dealing with solvent sensitive systems [8]. This is what we mean when we say "outliars can be considered upper/lower bounds"; a configuration in the centre of the ensemble average would be expected to obey the general trends, however those configurations found in this study do not. Therefore, they likely fall within the upper or lower limits of the ensemble average. To further investigate the reasonable nature of our initial conditions to optimisation (as this may effect the final result), the bulk configuration can be probed using molecular dynamics. Analysis of the center-of-mass radial distribution function $g(r)$ for solvent molecules relative to the xanthione solute (Figure 4a) shows the first solvation shell to manifest at around 2 Å, while beyond 8 Å, the system behaves like a bulk solution ($g(r) = 1$). The region between 2–8 Å acts as a buffer, or transition layer. Notably, at 4 Å, $g(r)$ deviates from the bulk value by less than 10%, indicating that the solvation process quickly approaches bulk behavior. At this distance, the number of nearest neighbour solvent molecules is approximately four (Figure 4b), true for acetone, toluene and DMSO. For water, a slower shift in the transition layer is found for the radial distribution function. However, the use of six water molecules remains a reasonable approximation for the solvation shell, correlating with previous observations of water in explicit solvent models [8]. This suggests that even with a relatively small number of solvent molecules, a realistic solvation process can be captured.

Interestingly, while $g(r)$ and the first solvation shell exhibit similar behaviour for all chosen solvents, their effects on xanthione differ significantly. The root mean square deviation (RMSD) analysis (Figure 4c) reveals two distinct peaks: the first corresponding to the implicit solvent geometry, and the second representing solvent-induced distortions. Among the solvents studied, toluene exhibited the largest deviation from its implicit solvent structure. As the dielectric constant of the solvent increases, RMSD decreases, indicating that solvent configurations deviate more from their gas-phase geometry in low-polarity environments. In contrast, solvents with higher dielectric constants lead to geometries more closely resembling the gas-phase structure. This suggests that low-polarity solvents may not fully sample the configuration space, instead favouring a limited set of atomic arrangements, which could introduce fluctuations in the results of quantum chemistry calculations if the implicit solvent model's configurations is close to its corresponding geometry. To address this, more accurate statistical ab initio methods should be incorporated in future work to improve the reliability of these simulations.

Examination of the radial distribution function of the heteroatoms (Figure 4d) shows that no strong interaction is observed between xanthione and DMSO. However, a strong solute-solvent interaction breaks the symmetry of the xanthione core and perturbs the equilibrium geometry compared to the symmetrical core observed in an implicit solvents; this interaction serves as a possible explanation for the explicit solvent results (Figure 3). No clear difference in the radial distribution function between acetone and DMSO is observable.

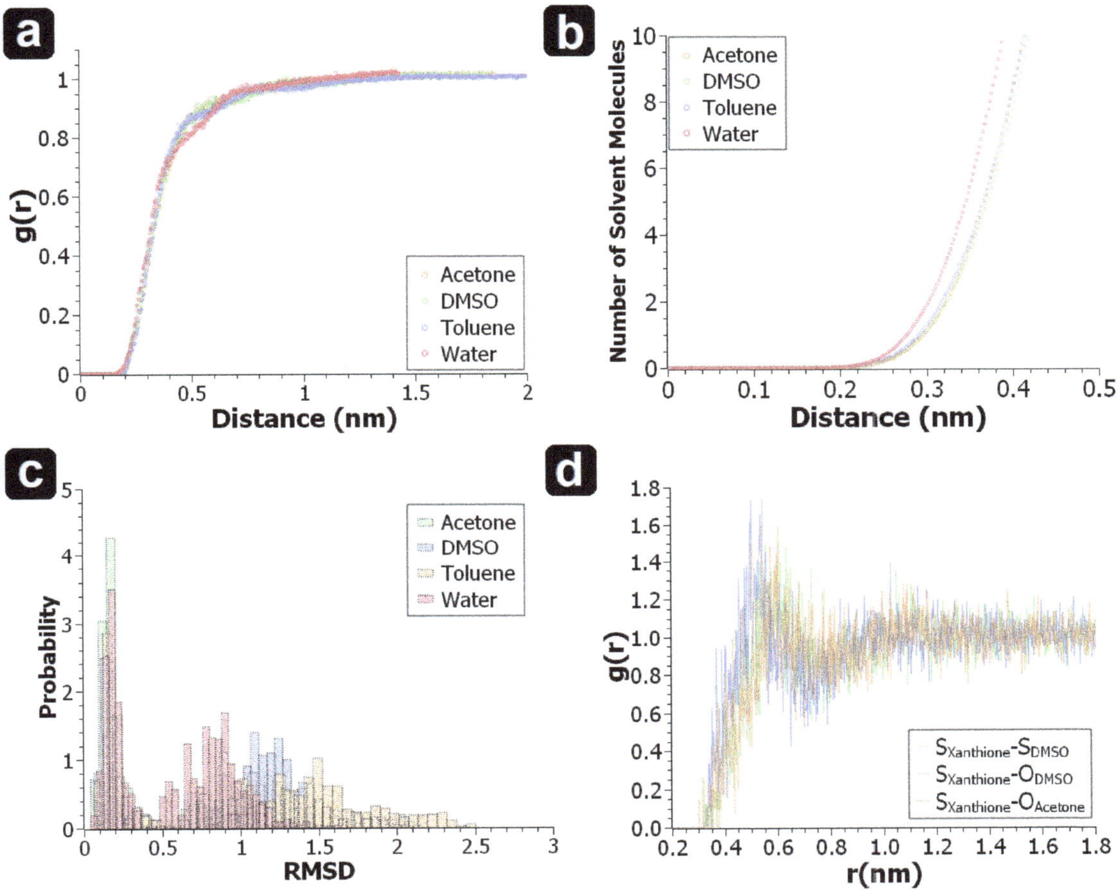

Figure 4. (**a**) Radial distribution function $g(r)$ of the center of mass of solvent molecules (acetone, dimethyl sulfoxide, toluene, and water) relative to xanthione. (**b**) Nearest neighbour solvent count as a function of distance from the solute. (**c**) Root mean square deviation (RMSD) distribution of xanthione over the MD trajectories in solutions with respect to the force field optimized structure in implicit solvent. (**d**) Comparison of the radial distribution function of the sulfur atom in xanthione in acetone and dimethyl sulfoxide.

Further on the solvents stabilisation; using an explicit model one would typically expect more of a pronounced effect between the solvent molecule and the sulfur-centre. However, while this work does capture the subtle environmental differences near the oxygen and nitrogen sites, it does not for the sulfur-centres in terms of the molecular configurations. Examination of the charge sites (−0.39 au, −0.35 au, and −0.41 au for S, O, and N, respectively) suggests that the contribution of the sulfur to overall dipole moment is not distinctive. That being said, even with similar solvent configurations it will likely still influence the electronic structure. Therefore, we acknowledge that while we observe no specific contribution due to the sulfur-centre, the electronic structure shows a clear dependence on explicit solvent interactions at this site.

4. Conclusions

This study provides a comprehensive investigation into the solvent effects on the xanthione chromophore, utilising both implicit and explicit solvent models. The use of implicit solvation via a PCM was found to significantly overestimate vertical excitation energies; this discrepancy was effectively corrected by incorporating explicit solvent molecules. While no inversion of the L_a and L_b states was detected, the L_a state consistently emerged as the lowest energy manifold in highly polar solvents. Molecular dynamics simulations further revealed that capturing the essential solvent-induced effects requires only a small number of explicit solvent molecules, with convergence achieved using 4–6 molecules for bulk properties, with less than a 10% error margin. Importantly, these results infer that energy levels as a function of solvent polarity trend towards a plateau once solvent polarity is beyond 60; while the L_b state is largely unaffected by this, the L_a state is shown to be highly sensitive when examining the adiabatic energy, while the L_b state is more sensitive in terms of vertical excitation energies. The energy gap is shown to be minimal for higher polarity mediums than lower.

This work lays a solid foundation for further exploration into the complex solvatochromic behaviour of xanthione, offering key insights into how solvent polarity influences its photophysical stability. The findings suggest that polar solvents can significantly stabilise xanthione's electronic states as well as its implicit solvent geometry, but also underscores the potential for further refinement in our understanding of solvent-induced modifications to its electronic structure. These results open avenues for future development of more precise quantum chemical models, which will be essential in harnessing xanthione's unique properties for applications in emerging quantum technologies.

Supplementary Materials: The following supporting information can be downloaded at: https://www.mdpi.com/article/10.3390/molecules29235609/s1, Electronic supplementary information: Simulation of solvatochromic phenomena in xanthione using explicit solvent methods.

Author Contributions: Conceptualisation, A.M., Z.C. and R.J.H.; methodology, A.M. and Z.C.; software, A.M. and Z.C.; validation, A.M., Z.C. and R.J.H.; formal analysis, A.M. and Z.C.; investigation, A.M. and Z.C.; resources, S.P.R.; data curation, A.M. and Z.C.; writing—original draft preparation, A.M., Z.C. and R.J.H.; writing—review and editing, A.M. and Z.C.; visualization, A.M. and Z.C.; supervision, S.P.R.; project administration, S.P.R.; funding acquisition, S.P.R. All authors have read and agreed to the published version of the manuscript.

Funding: This work was supported by the Australian Government through the Australian Research Council (ARC) under the Centre of Excellence scheme (project number CE170100026).

Institutional Review Board Statement: Not applicable.

Informed Consent Statement: Not applicable.

Data Availability Statement: Electronic Supplementary Information contains all absolute energies (hartree) for xanthione in each solvent for implicit and explicit solvent configurations, rendered frontier molecular orbitals, and the optimised coordinates for each configurations studied in this work.

Acknowledgments: A.M. thanks Wallace W. H. Wong for advise on chemistry related to this work.

Conflicts of Interest: The authors declare no conflict of interest.

References

1. Rai-Constapel, V.; Etinski, M.; Marian, C.M. Photophysics of Xanthone: A Quantum Chemical Perusal. *J. Phys. Chem. A* **2013**, *117*, 3935–3944. [CrossRef] [PubMed]
2. Bondarev, S.L.; Tikhomirov, S.A.; Buganov, O.V.; Knyukshto, V.N.; Galinovskii, N.A.; Fedunov, R.G.; Khokhlova, S.S.; Ivanov, A.I. Spectral Dynamics of Nitro Derivatives of Xanthione in Solutions. *J. Phys. Chem. A* **2019**, *123*, 1570–1580. [CrossRef] [PubMed]
3. Steer, R.P. Structure and decay dynamics of electronic excited states of thiocarbonyl compounds. *Rev. Chem. Intermed.* **1981**, *4*, 1–41. [CrossRef]
4. Mundt, R.; Villnow, T.; Ziegenbein, C.T.; Gilch, P.; Marian, C.; Rai-Constapel, V. Thioxanthone in apolar solvents: Ultrafast internal conversion precedes fast intersystem crossing. *Phys. Chem. Chem. Phys.* **2016**, *18*, 6637–6647. [CrossRef]
5. Huang, C.H.; Wu, C.C.; Li, E.Y.; Chou, P.T. Quest for singlet fission of organic sulfur-containing systems in the higher lying singlet excited state: Application prospects of anti-Kasha's rule. *Phys. Chem. Chem. Phys.* **2023**, *25*, 9115–9122. [CrossRef]
6. Robert Huber, J.; Mahaney, M. $S_2 \rightarrow S_0$ fluorescence in an aromatic thioketone, xanthione. *Chem. Phys. Lett.* **1975**, *30*, 410–412. [CrossRef]

7. Mahaney, M.; Huber, J. Fluorescence from the second excited singlet of aromatic thioketones in solution. *Chem. Phys.* **1975**, *9*, 371–378. [CrossRef]
8. Manian, A.; Shaw, R.A.; Lyskov, I.; Russo, S.P. The quantum chemical solvation of indole: Accounting for strong solute–solvent interactions using implicit/explicit models. *Phys. Chem. Chem. Phys.* **2022**, *24*, 3357–3369. [CrossRef]
9. Abou-Hatab, S.; Carnevale, V.; Matsika, S. Modeling solvation effects on absorption and fluorescence spectra of indole in aqueous solution. *J. Chem. Phys.* **2021**, *154*, 064104. [CrossRef]
10. Xin, X.; Shi, W.; Jia, R.; Zhao, G.; Zhuang, H.; Li, Y. A theoretical elucidation of the role of explicit solvent in the excited-state double proton transfer process for 2,4-Bisimidazolylphenol. *J. Mol. Struct.* **2024**, *1318*, 139328. [CrossRef]
11. Pervin, R.; Manian, A.; Chen, Z.; Christofferson, A.J.; Owyong, T.C.; Bradley, S.J.; White, J.M.; Ghiggino, K.P.; Russo, S.P.; Wong, W.W. Medium effects on the fluorescence of Imide-substituted naphthalene diimides. *J. Photochem. Photobiol. A Chem.* **2023**, *436*, 114364. [CrossRef]
12. Liu, J.J.; Xia, S.B.; Liu, D.; Hou, J.; Suo, H.; Cheng, F.X. Multifunctional naphthalene diimide-based coordination polymers: Photochromism and solventchromism. *Dye. Pigment.* **2020**, *177*, 108269. [CrossRef]
13. Manian, A.; Campaioli, F.; Hudson, R.J.; Cole, J.H.; Schmidt, T.W.; Lyskov, I.; Smith, T.A.; Russo, S.P. Charge Transfer-Mediated Multi-exciton Mechanisms in Weakly Coupled Perylene Dimers. *Chem. Mater.* **2023**, *35*, 6889–6908. [CrossRef]
14. Platt, J.R. Classification of Spectra of Cata-Condensed Hydrocarbons. *J. Chem. Phys.* **1949**, *17*, 484–495. [CrossRef]
15. Becke, A.D. Density-functional thermochemistry. III. The role of exact exchange. *J. Chem. Phys.* **1993**, *98*, 5648–5652. [CrossRef]
16. Lee, C.; Yang, W.; Parr, R.G. Development of the Colle-Salvetti correlation-energy formula into a functional of the electron density. *Phys. Rev. B* **1988**, *37*, 785–789. [CrossRef]
17. Vosko, S.H.; Wilk, L.; Nusair, M. Accurate spin-dependent electron liquid correlation energies for local spin density calculations: A critical analysis. *Can. J. Phys.* **1980**, *58*, 1200–1211. [CrossRef]
18. Stephens, P.J.; Devlin, F.J.; Chabalowski, C.F.; Frisch, M.J. Ab Initio Calculation of Vibrational Absorption and Circular Dichroism Spectra Using Density Functional Force Fields. *J. Phys. Chem.* **1994**, *98*, 11623–11627. [CrossRef]
19. Yanai, T.; Tew, D.P.; Handy, N.C. A new hybrid exchange–correlation functional using the Coulomb-attenuating method (CAM-B3LYP). *Chem. Phys. Lett.* **2004**, *393*, 51–57. [CrossRef]
20. Grimme, S.; Antony, J.; Ehrlich, S.; Krieg, H. A consistent and accurate ab initio parametrization of density functional dispersion correction (DFT-D) for the 94 elements H-Pu. *J. Chem. Phys.* **2010**, *132*, 154104. [CrossRef]
21. Grimme, S.; Ehrlich, S.; Goerigk, L. Effect of the damping function in dispersion corrected density functional theory. *J. Comput. Chem.* **2011**, *32*, 1456–1465. [CrossRef] [PubMed]
22. Ditchfield, R.; Hehre, W.J.; Pople, J.A. Self-Consistent Molecular-Orbital Methods. IX. An Extended Gaussian-Type Basis for Molecular-Orbital Studies of Organic Molecules. *J. Chem. Phys.* **1971**, *54*, 724–728. [CrossRef]
23. Hehre, W.J.; Ditchfield, R.; Pople, J.A. Self—Consistent Molecular Orbital Methods. XII. Further Extensions of Gaussian—Type Basis Sets for Use in Molecular Orbital Studies of Organic Molecules. *J. Chem. Phys.* **1972**, *56*, 2257–2261. [CrossRef]
24. Hariharan, P.C.; Pople, J.A. The influence of polarization functions on molecular orbital hydrogenation energies. *Theor. Chim. Acta* **1973**, *28*, 213–222. [CrossRef]
25. Frisch, M.J.; Trucks, G.W.; Schlegel, H.B.; Scuseria, G.E.; Robb, M.A.; Cheeseman, J.R.; Scalmani, G.; Barone, V.; Petersson, G.A.; Nakatsuji, H.; et al. *Gaussian16 Revision B.01*; Gaussian Inc.: Wallingford, CT, USA, 2016.
26. Grimme, S.; Waletzke, M. A combination of Kohn–Sham density functional theory and multi-reference configuration interaction methods. *J. Chem. Phys.* **1999**, *111*, 5645–5655. [CrossRef]
27. Lyskov, I.; Kleinschmidt, M.; Marian, C.M. Redesign of the DFT/MRCI Hamiltonian. *J. Chem. Phys.* **2016**, *144*, 034104. [CrossRef] [PubMed]
28. Becke, A.D. Density-functional exchange-energy approximation with correct asymptotic behavior. *Phys. Rev. A* **1988**, *38*, 3098–3100. [CrossRef]
29. Becke, A.D. A new mixing of Hartree–Fock and local density-functional theories. *J. Chem. Phys.* **1993**, *98*, 1372–1377. [CrossRef]
30. Weigend, F.; Ahlrichs, R. Balanced basis sets of split valence, triple zeta valence and quadruple zeta valence quality for H to Rn: Design and assessment of accuracy. *Phys. Chem. Chem. Phys.* **2005**, *7*, 3297–3305. [CrossRef]
31. Schäfer, A.; Huber, C.; Ahlrichs, R. Fully optimized contracted Gaussian basis sets of triple zeta valence quality for atoms Li to Kr. *J. Chem. Phys.* **1994**, *100*, 5829–5835. [CrossRef]
32. University of Karlsruhe. TURBOMOLE V7.3 2018, a Development of University of Karlsruhe and Forschungszentrum Karlsruhe GmbH, TURBOMOLE GmbH, Since 2007; 1989–2007. Available online: https://www.turbomole.org/ (accessed on 1 January 2020).
33. Abraham, M.J.; Murtola, T.; Schulz, R.; Páll, S.; Smith, J.C.; Hess, B.; Lindahl, E. GROMACS: High performance molecular simulations through multi-level parallelism from laptops to supercomputers. *SoftwareX* **2015**, *1–2*, 19–25. [CrossRef]
34. Wang, J.; Wolf, R.M.; Caldwell, J.W.; Kollman, P.A.; Case, D.A. Development and testing of a general amber force field. *J. Comput. Chem.* **2004**, *25*, 1157–1174. [CrossRef] [PubMed]
35. Bayly, C.I.; Cieplak, P.; Cornell, W.; Kollman, P.A. A well-behaved electrostatic potential based method using charge restraints for deriving atomic charges: The RESP model. *J. Org. Chem.* **1993**, *97*, 10269–10280. [CrossRef]
36. Cornell, W.D.; Cieplak, P.; Bayly, C.I.; Kollman, P.A. Application of RESP charges to calculate conformational energies, hydrogen bond energies, and free energies of solvation. *J. Am. Chem. Soc.* **1993**, *115*, 9620–9631. [CrossRef]

37. Schäfer, A.; Horn, H.; Ahlrichs, R. Fully optimized contracted Gaussian basis sets for atoms Li to Kr. *J. Chem. Phys.* **1992**, *97*, 2571–2577. [CrossRef]
38. Darden, T.; York, D.; Pedersen, L. Particle mesh Ewald: An N*log*(N) method for Ewald sums in large systems. *J. Chem. Phys.* **1993**, *98*, 10089–10092. [CrossRef]
39. Nosé, S. A unified formulation of the constant temperature molecular dynamics methods. *J. Chem. Phys.* **1984**, *81*, 511–519. [CrossRef]
40. Hoover, W.G. Canonical dynamics: Equilibrium phase-space distributions. *Phys. Rev. A* **1985**, *31*, 1695–1697. [CrossRef]
41. Berendsen, H.J.C.; Postma, J.P.M.; van Gunsteren, W.F.; DiNola, A.; Haak, J.R. Molecular dynamics with coupling to an external bath. *J. Chem. Phys.* **1984**, *81*, 3684–3690. [CrossRef]
42. Capitanio, D.; Pownall, H.; Huber, J. Spectroscopic and photochemical properties of aromatic thioketones: xanthione. *J. Photochem.* **1974**, *3*, 225–236. [CrossRef]
43. Riddick, J.A.; Bunger, W.B.; Sakano, T.K. *Organic Solvents: Physical Properties and Methods of Purification*; Wiley: Hoboken, NJ, USA, 1986; p. 1325.
44. Lorenc, M.; Maciejewski, A.; Ziolek, M.; Naskrecki, R.; Karolczak, J.; Kubicki, J.; Ciesielska, B. Mechanism and deactivation kinetics of S_2-xanthione in acetonitrile, a quenching solvent, and of S_2-exciplex measured by pico- and femtosecond laser spectroscopy. *Chem. Phys. Lett.* **2001**, *346*, 224–232. [CrossRef]
45. Maciejewski, A.; Steer, R.P. Photophysics of the second excited singlet states of xanthione and related thiones in perfluoroalkane solvents. *J. Am. Chem. Soc.* **1983**, *105*, 6738–6740. [CrossRef]
46. Banks, R.E.; Smart, B.E.; Tatlow, J.C. (Eds.) *Organofluorine Chemistry: Principles and Commercial Applications*; Springer: New York, NY, USA, 1994. [CrossRef]
47. Maciejewski, A.; Jakubowska, A.; Dutkiewicz, E.; Augustyniak, W. Spectral and Photophysical Properties of Aromatic Thiones in Micellar Systems: Xanthione as Convenient Absorption and Emission Probes. *J. Colloid Interface Sci.* **1996**, *177*, 528–541. [CrossRef]

molecules

MDPI

Article

Population and Energy Transfer Dynamics in an Open Excitonic Quantum Battery

Zhe Liu and Gabriel Hanna *

Department of Chemistry, University of Alberta, Edmonton, AB T6G 2G2, Canada; zhe17@ualberta.ca

* Correspondence: gabriel.hanna@ualberta.ca

Abstract: In a previous study, we proposed an open quantum network model of a quantum battery (QB) that possesses dark states owing to its structural exchange symmetries. While in a dark state, the QB is capable of storing an exciton without any environment-induced population losses. However, when the structural exchange symmetry is broken, the QB begins to discharge the exciton towards its exit site. In this article, we start by demonstrating that this QB is not only loss-free with respect to exciton population during the storage phase, but also with respect to the QB energy. We then explore the exciton population and energy transfer dynamics of the QB during the discharge phase over a wide range of site energies, bath temperatures, and bath reorganization energies. Our results shed light on how to optimize the QB's population and energy transfer dynamics for different purposes.

Keywords: quantum battery; exciton dynamics; quantum energy storage; dark state

1. Introduction

A quantum battery (QB) is a quantum system that can store and release energy as needed. Over the past decade, several types of QBs have been proposed, demonstrating advantages with respect to charging speed [1–9] and work extraction [5,7,10–14] over the classical analogues. In practice, a QB may interact with an environment, which could lead to losses that negatively impact its performance. Thus, any theoretical study should treat the QB as an open quantum system [15–17]. Previously, open quantum network (OQN) models, in which the network sites (representing quantum systems) may be coupled to each other and to dissipative/decohering environments, have been used to study the dynamics of open QBs [5,6,8,18].

Over the years, a number of ways of protecting a quantum system from its environment have been proposed. One well-known approach is based on the use of a decoherence-free subspace (DFS) [19,20]—a subspace of the Hilbert space in which the dynamics is purely unitary. While in a DFS, the system's dynamics is dissipationless and decoherence-free despite the coupling to an environment. As has been shown in Refs. [21–23], if there exists a unitary operator that commutes with all elements in the system's master equation exists, then the system will possess invariant subspaces. Among such subspaces, the one-dimensional subspaces are DFSs because the dynamics maps them onto themselves. Owing to this property, DFSs have many potential applications in quantum information and quantum computing [24–29].

Recently, researchers have used dark states living in DFSs for stabilizing and enhancing the performance of QBs and other OQNs [18,30–32]. In particular, we proposed an OQN model of a QB with site exchange symmetries that support the existence of dark states localized on the bulk sites of the network [18]. We showed that it is possible to store an exciton in one of these dark states, without any exciton population transfer to the two surface sites which are connected to thermal baths at equal temperatures. In this way, it was possible to protect the QB from environment-induced excitation energy losses. Moreover, we showed that by attaching an additional bath to break the exchange symmetry of the

QB, it is possible to discharge the exciton towards the designated exit site to be ultimately harnessed by a sink.

Although it was shown that our QB model is capable of operating as an *excitonic* QB (i.e., a QB that stores and discharges excitons), the question of whether or not the model is capable of operating as an *energy* battery (i.e., a QB whose energy does not decrease significantly during the storage phase and discharges energy to a sink with minimal loss to the baths) remained to be explored. Therefore, in this study, in addition to monitoring the exciton population dynamics, we calculate the various contributions to the total energy of the system during both the storage and discharge phases for a wide range of bath temperature gaps, bath reorganization energies, and site energies. The aim of the study is to identify the conditions which optimize the performance of the QB model for its use as an excitonic QB, energy QB, or both.

The paper is organized as follows. First, in Section 2, we introduce the open QB model. In Section 3, we explain how the dynamics of the model are simulated and how the populations/energies are calculated. In Section 4, we present the time-dependent populations/energies of the model for different bath temperatures, bath reorganization energies, and site energies. The performance of the QB under the various conditions is also discussed. In Section 5, we summarize our findings.

2. Open QB Model

Following Ref. [18], we consider the same six-site para-benzene-shaped OQN (the numbering of the sites is depicted in Figure 1b). Setting $\hbar = 1$, the Hamiltonian of the closed network is given by

$$\hat{H}_N = \sum_{n=1}^{6} E_n |n\rangle\langle n| + h \sum_{\langle n,m\rangle} |n\rangle\langle m|, \tag{1}$$

where $|n\rangle$ corresponds to a singly excited state localized on site n, E_n is the energy of site n, h is the electronic coupling strength between sites n and m, and $\langle n, m\rangle$ denotes that a cyclic summation over nearest-neighbour sites is performed. To construct the OQN, the two para-sites of the network are coupled to thermal baths, each composed of a set of independent harmonic oscillators. The sum of the bath and network–bath coupling Hamiltonians is given by

$$\hat{H}_B + \hat{H}_{NB} = \frac{1}{2} \sum_{n \in SSs} \sum_{j}^{M} \left[\hat{P}_{n,j}^2 + \omega_{n,j}^2 \left(\hat{R}_{n,j} - \frac{C_{n,j}}{\omega_{n,j}^2} |n\rangle\langle n| \right)^2 \right], \tag{2}$$

where M is the number of harmonic oscillators in each bath, $\hat{P}_{n,j}$ and $\hat{R}_{n,j}$ are the mass-weighted momentum and position operators, respectively, of the jth oscillator with frequency $\omega_{n,j}$, and $C_{n,j}$ is the network–bath coupling strength between the nth site and the jth oscillator. The sites coupled to the baths are referred to as surface sites (SSs) while the remaining sites are referred to as bulk sites (BSs). In this study, we take the site energies of all BS to be equal to a value E_{BS}, while considering different values of site energies of the SS.

The network defined by $\hat{H}_N$ and the aforementioned site energies possesses the following unitary symmetry operator [33]

$$\hat{\Pi} = |1\rangle\langle 1| + |4\rangle\langle 4| + |2\rangle\langle 6| + |3\rangle\langle 5| + h.c., \tag{3}$$

which satisfies

$$[\hat{\Pi}, \hat{H}_N] = 0, \quad [\hat{\Pi}, \hat{H}_{NB}] = 0 \quad \forall n \in \text{SSs}. \tag{4}$$

As a result, $\hat{\Pi}$ shares the same eigenstates as $\hat{H}_N$. Due to the existence of this symmetry operator, the system possesses two DFSs with the following DSs ($|\psi_\alpha\rangle$) and eigenvalues (u_α) [22,34]

$$\begin{aligned} |\psi_1\rangle &= \frac{1}{2}(|5\rangle + |6\rangle - |2\rangle - |3\rangle), \quad u_1 = E_{BS} + h \\ |\psi_2\rangle &= \frac{1}{2}(|3\rangle + |6\rangle - |2\rangle - |5\rangle), \quad u_2 = E_{BS} - h \end{aligned} \tag{5}$$

If the system is initialized in the dark state $|\psi_\alpha\rangle$, it will undergo dissipationless dynamics, i.e., the dark state will be invariant under the effect of the evolution operator of the composite system and the site populations will be

$$\langle \hat{\mathcal{P}}_{nn}(t)\rangle = \langle \hat{\mathcal{P}}_{nn}(0)\rangle = \begin{cases} 0, & \forall n \in \text{SSs}, \\ \frac{1}{4}, & \forall n \in \text{BSs}, \end{cases} \tag{6}$$

where $\hat{\mathcal{P}}_{nn} = |n\rangle\langle n|$ is the projection operator corresponding to site n and $\langle\cdot\rangle$ denotes an ensemble average. Because an exciton can be stored indefinitely in this state, this phase is termed the *storage phase*.

When a symmetry-breaking perturbation (SBP) is connected to the OQN, the symmetry operator and DSs no longer persist and the QB may begin to discharge a stored exciton. In this study, this is achieved by attaching a bath of M harmonic oscillators to sites 2 and 3 simultaneously. The sum of the SBP and network-SBP coupling Hamiltonians is

$$\hat{H}_P + \hat{H}_{NP} = \frac{1}{2}\sum_k^M \left[\hat{p}_k^2 + \Omega_k^2\left(\hat{r}_k - \frac{\gamma_k}{\Omega_k^2}\hat{S}\right)^2\right], \tag{7}$$

where $\hat{p}_k$, $\hat{r}_k$, Ω_k, and γ_k are the momentum operator, position operator, frequency, and network–SBP coupling strength of the kth oscillator, respectively, and $\hat{S} = |2\rangle\langle 2| + |3\rangle\langle 3|$. This phase is termed the *discharge phase*.

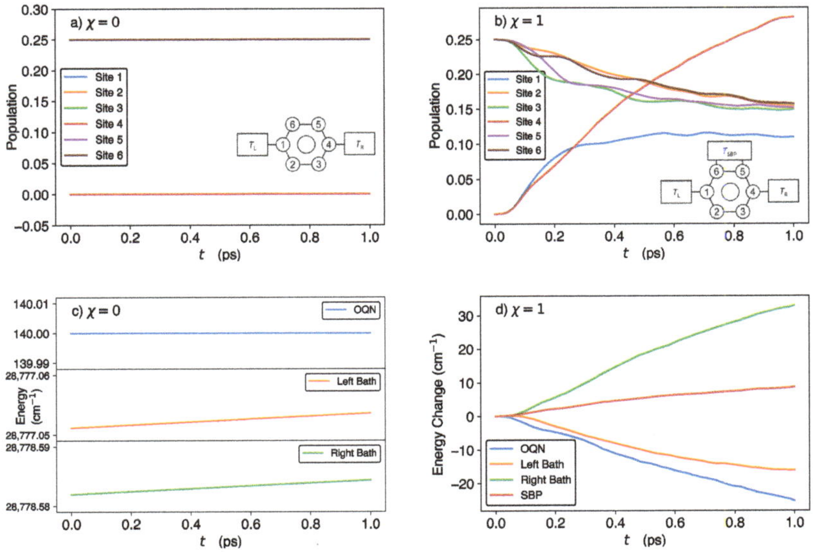

Figure 1. Time-dependent site populations (**upper panels**) and energies (**lower panels**) of the OQN, SBP, and baths in the storage (**left panels**) and discharge (**right panels**) phases. In panel (**d**),

energy changes are calculated by subtracting the initial value of the energy from the value at each time. The results were generated using the following parameter set: $E_1 = 250\ \text{cm}^{-1}$, $E_{i\in\{2,3,5,6\}} = 200\ \text{cm}^{-1}$, $E_4 = 0\ \text{cm}^{-1}$, $h = -60\ \text{cm}^{-1}$, $T_L = T_R = 300$ K, $T_p = 300$ K, $\lambda_b = 35\ \text{cm}^{-1}$, $\lambda_p = 10\ \text{cm}^{-1}$, $\omega_c = \omega_p = 106\ \text{cm}^{-1}$, $\omega_{max} = 50\omega_c$, and $M = 100$.

3. Simulation Details

Due to the large number of degrees of freedom in the composite system (i.e., QB, thermal baths, and SBP), a fully quantum dynamical simulation of the composite system would be computationally expensive. Thus, following Ref. [18], we use a mixed quantum–classical dynamics method known as "Deterministic evolution of coordinates with initial decoupled equations" (DECIDE) [35,36], which treats the OQN quantum mechanically and the baths and SBP in a classical-like way. Previously, the DECIDE method has been successfully applied to a host of model systems over a large range of parameter regimes [35,37–40], and is, therefore, expected to produce reliable results in this study. That being said, DECIDE may yield inaccurate results for systems with very slow thermal baths (i.e., when the bath cut-off frequency is much smaller than the subsystem energy gaps) or with very low bath temperatures, neither of which is the case in the present study.

To apply the DECIDE method, we must first apply the Wigner transform [41] to the bath and SBP degrees of freedom. The resulting partially Wigner-transformed Hamiltonian of the composite system is

$$\hat{H}_W = \hat{H}_N + \hat{H}_{NB}(\{R_{n,j}\}) + H_B(\{P_{n,j}\},\{R_{n,j}\}) + \chi\left[H_P(\{p_k\},\{r_k\}) + \hat{H}_{NP}(\{r_k\})\right], \quad (8)$$

where $\{R_{n,j}, P_{n,j}\}$ and $\{r_k, p_k\}$ are the position and momentum variables of the baths and SBP, respectively. The parameter χ is equal to 1 when the SBP is attached to the OQN and 0 otherwise. The coordinates of the OQN are taken to be $\hat{\mathcal{P}}_{nm} = |n\rangle\langle m|$, while the coordinates of the baths and SBP are their positions and momenta. According to the DECIDE method, the coupled equations of motion for all of the coordinates are given by [18]

$$\begin{aligned}
\frac{d}{dt}\mathcal{P}_{nm}^{\beta\beta'}(t) &= i\left[\sum_{l=1}^{6} V_{ln}\hat{\mathcal{P}}_{lm}(t) - \sum_{v=1}^{6} V_{mv}\hat{\mathcal{P}}_{nv}(t)\right]^{\beta\beta'} \\
&\quad -\frac{i}{2}\sum_j C_{n,j}(R_{n,j}(t)\hat{\mathcal{P}}_{nm}(t) + \hat{\mathcal{P}}_{nm}(t)R_{n,j}(t))^{\beta\beta'}(\delta_{n,1}+\delta_{n,4}) \\
&\quad +\frac{i}{2}\sum_j C_{m,j}(R_{m,j}(t)\hat{\mathcal{P}}_{nm}(t) + \hat{\mathcal{P}}_{nm}(t)R_{m,j}(t))^{\beta\beta'}(\delta_{m,1}+\delta_{m,4}) \\
&\quad -\chi\frac{i}{2}\sum_k C_k(R_k(t)\hat{\mathcal{P}}_{nm}(t) + \hat{\mathcal{P}}_{nm}(t)R_k(t))^{\beta\beta'}(\delta_{n,2}+\delta_{n,3}) \\
&\quad +\chi\frac{i}{2}\sum_k C_k(R_k(t)\hat{\mathcal{P}}_{nm}(t) + \hat{\mathcal{P}}_{nm}(t)R_k(t))^{\beta\beta'}(\delta_{m,2}+\delta_{m,3}), \\
\frac{d}{dt}R_{n,j}^{\beta\beta'}(t) &= P_{n,j}^{\beta\beta'}(t), \\
\frac{d}{dt}P_{n,j}^{\beta\beta'}(t) &= -\omega_{n,j}^2 R_{n,j}^{\beta\beta'}(t) + C_{n,j}\mathcal{P}_{nn}^{\beta\beta'}(t)(\delta_{n,1}+\delta_{n,4}), \\
\frac{d}{dt}r_k^{\beta\beta'}(t) &= p_k^{\beta\beta'}(t), \\
\frac{d}{dt}p_k^{\beta\beta'}(t) &= -\Omega_k^2 r_k^{\beta\beta'}(t) + \chi\gamma_k(\hat{\mathcal{P}}_{22}+\hat{\mathcal{P}}_{33})^{\beta\beta'}(t),
\end{aligned} \quad (9)$$

where $V_{nn} = E_n + \sum_{j=1}^{M} C_{n,j}^2/(2\omega_{n,j}^2)(\delta_{n,1}+\delta_{n,4}) + \chi\sum_{k=1}^{M}\gamma_k^2/(2\Omega_k^2)(\delta_{n,2}+\delta_{n,3})$, and when $n \neq m$, $V_{nm} = h$ for $|n-m| = 1$ and 0 otherwise. Here, β labels an arbitrary basis state, i.e., the matrix element of $\hat{\mathcal{P}}_{nm}$ is given by $\mathcal{P}_{nm}^{\beta\beta'} = \langle\beta|\hat{\mathcal{P}}_{nm}|\beta'\rangle$. In this work, $\{|\beta\rangle\} = \{|1\rangle, |2\rangle, \ldots, |6\rangle\}$.

We assume the initial state of the composite system to be factorized, i.e., $\hat{\rho}_{tot}(0) = \hat{\rho}_N(0)\rho_{B,W}(0)\rho_{P,W}(0)$, where $\hat{\rho}_N(0)$ is the initial density operator of the network, and $\rho_{B,W}(0)$ and $\rho_{P,W}(0)$ are the initial Wigner-transformed densities of the heat baths and SBP, respectively (N.B.: $\rho_{P,W}(0)$ is omitted when the OQN is not attached to the SBP). The initial state of the network is taken to be the dark state $\hat{\rho}_N(0) = |\psi_1\rangle\langle\psi_1|$, where $|\psi_1\rangle$ is defined in Equation (5). The initial values of the OQN coordinates are always taken to be $\mathcal{P}_{nm}^{\beta\beta'} = \delta_{\beta,n}\delta_{m,\beta'}$. The bath oscillators are initialized in the thermal equilibrium state given by (setting $k_B = 1$) [42]

$$\rho_{B,W}(0) = \prod_{n=1,4}\prod_{j=1}^{M}\frac{\tanh(\beta\omega_{n,j}/2)}{\pi}\exp\left[-\frac{2\tanh(\beta\omega_{n,j}/2)}{\omega_{n,j}}\left(\frac{P_{n,j}^2}{2}+\frac{\omega_{n,j}^2R_{n,j}^2}{2}\right)\right], \tag{10}$$

where $\beta = 1/k_BT$ is the inverse temperature. The oscillators of the SBP are also initialized in a thermal equilibrium state with an analogous form. The initial positions and momenta of the bath and SBP oscillators are sampled from Equation (10) and its analog for the SBP, respectively. The system–bath and system–SBP couplings are characterized by a Debye–Drude spectral density, i.e., $J(\omega_{n,j}) = 2\lambda_b\frac{\omega_{n,j}}{\omega_{n,j}^2+1}$. In this work, the spectral density is discretized to yield the following expressions for the coupling strengths $C_{n,j}$ and frequencies $\omega_{n,j}$ [43,44]:

$$\omega_{n,j} = \tan(j\arctan(\omega_{max}/\omega_c)/M)\omega_c, \tag{11}$$

$$C_{n,j} = 2\sqrt{\lambda_b\arctan(\omega_{max}/\omega_c)/(\pi M)}\omega_{n,j}, \tag{12}$$

where λ_b is the bath reorganization energy and ω_c is the bath cut-off frequency [45].

Previously, the fourth-order Runge–Kutta method was used to integrate the DECIDE equations of motion in Equation (9), yielding conserved total populations for the system under study [18,35]. However, in this work, we found that a combination of a smaller time step and a higher order integrator is needed for good energy conservation and more accurate calculations of the various contributions to the total energy. High-order methods such as the eighth-order Runge–Kutta method [46] can yield accurate results with a relatively large time step, but it contains many integration stages which increase the simulation time drastically. Considering the trade-off between time step and number of integration steps, we employed the sixth-order Runge–Kutta method in this work. Using this integrator with a time step of 0.16 fs, the total energy drift is less than 10^{-2} cm^{-1} over a 1 ps trajectory [see Section S1 of the Supporting Information (SI)].

The time-dependent population of site n is calculated via an ensemble average of the projection operator $\hat{P}_{nn}$, viz.,

$$\langle\hat{\mathcal{P}}_{nn}(t)\rangle = \sum_{\beta\beta'}\int d\mathbf{X}(0)\mathcal{P}_{nn}^{\beta\beta'}(t)\rho_N^{\beta'\beta}(0)\rho_{E,W}(0), \tag{13}$$

where $\mathbf{X(0)} = (\{R_{n,j}\},\{P_{n,j}\},\{r_k\},\{p_k\})$ are the initial coordinates of the baths and SBP. Similarly, the average total energy of the composite system is

$$\langle E_{tot}(t)\rangle = \sum_{\beta\beta'}\int d\mathbf{X}(0)\hat{H}_W^{\beta\beta'}(t)\rho_N^{\beta'\beta}(0)\rho_{E,W}(0). \tag{14}$$

In the above equations, $\rho_{E,W} = \rho_{B,W}$ in the absence of the SBP and $\rho_{E,W} = \rho_{B,W}\rho_{P,W}$ when the SBP is attached to the QB. Using our numerical results, we have verified that $\sum_{n=1}^{6}\langle\hat{\mathcal{P}}_{nn}(t)\rangle \approx 1$ (i.e., population conservation) and $\frac{d}{dt}\langle E_{tot}(t)\rangle \approx 0$ (i.e., energy conservation). For the purposes of our analysis, the total energy of the composite system may be decomposed into the following contributions: OQN energy (E_N), bath energy ($E_{NB} + E_B$), and SBP energy ($E_{NP} + E_P$). All simulation results are averaged over 10,000 trajectories, which ensures that the error bars are much smaller than the symbols in the figures.

4. Results and Discussion

4.1. Exciton Storage and Discharge

We start by considering the time-dependent site populations and energies of the OQN, baths, and SBP in the storage ($\chi = 0$) and discharge ($\chi = 1$) phases, using the parameter values from Ref. [18], viz., $E_1 = 250$ cm^{-1}, $E_{i\in\{2,3,5,6\}} = 200$ cm^{-1}, $E_4 = 0$ cm^{-1} (N.B.: E_4 is smaller than the energies of the remaining sites because site 4 is designated as the exit site, which in practice would be attached to a sink that captures the exciton/energy), $h = -60$ cm^{-1}, $T_L = T_R = 300$ K, $T_p = 300$ K, $\lambda_b = 35$ cm^{-1}, $\lambda_p = 10$ cm^{-1}, $\omega_c = \omega_p = 106$ cm^{-1}, $\omega_{max} = 50\omega_c$, and $M = 100$.

In Figure 1a, we see that the BS and SS populations remain constant at 0.25 and 0, respectively, during the storage phase. Simultaneously, as seen in Figure 1c, the OQN, left-bath, and right-bath energies remain constant (the minor deviations in the bath energies are attributed to numerical errors). Thus, our QB model is capable of perfectly storing both population and energy during the storage phase, as predicted by the theory. In addition, as shown in Section S3 of the SI, the coherences remain constant during the storage phase.

After attaching the SBP, the populations of sites 1 and 4 increase from 0 while the remaining site populations decrease from 0.25 over the 1 ps time period, as seen in Figure 1b. At $t = 1$ ps, the population of site 4 (i.e., the exit site) is greater than that of site 1 and the BS. (The corresponding time-dependent coherences are shown in Figure S2 of the SI). In Figure 1d, we plot the changes in the OQN, left-bath, right-bath, and SBP energies with respect to their initial values. As can be seen, the OQN and left-bath energies decrease while the SBP and right-bath energies increase. Thus, during the discharge phase, energy flows from the OQN and left bath into the right bath and SBP, with considerably more energy flowing to the right bath than to the SBP.

To further analyze the OQN energy (E_N) change in the discharge phase, we decompose it into the on-site energy and exchange energy, corresponding to the terms $\sum_{n=1}^{6} E_n |n\rangle\langle n|$ and $h\sum_{\langle n,m\rangle} |n\rangle\langle m|$, respectively, in Equation (1). The change in the exchange energy is therefore given by

$$\Delta E_{\text{exch}}(t) = h \sum_{\langle n,m\rangle} \left[\langle \hat{\mathcal{P}}_{nm}(t)\rangle - \langle \hat{\mathcal{P}}_{nm}(0)\rangle\right]. \tag{15}$$

At $t = 0$, only $\langle \hat{\mathcal{P}}_{23}\rangle$, $\langle \hat{\mathcal{P}}_{32}\rangle$, $\langle \hat{\mathcal{P}}_{56}\rangle$, and $\langle \hat{\mathcal{P}}_{65}\rangle$ are non-zero (since the OQN is initialized in the dark state $|\psi_1\rangle$). Their values decrease from 0.25 to ≈ 0.07 over the course of 1 ps (Figure S4), which causes ΔE_{exch} to increase. For the dynamics displayed in Figure 1b,d, ΔE_{exch} increases by 25.59 cm^{-1} over the course of 1 ps (as calculated by Equation (15)). In fact, ΔE_{exch} increases for all of the parameter regimes studied herein. Considering the values of $\langle \hat{\mathcal{P}}_{nn}(0)\rangle$ and the fact that $\sum_{n=1}^{6} \langle \hat{\mathcal{P}}_{nn}(t)\rangle = 1$, the on-site energy change is (setting all BS energies to be equal, i.e., $E_{i\in\{2,3,5,6\}} = E_{BS}$)

$$\begin{aligned} \Delta E_{\text{on-site}}(t) &= \sum_{n=1}^{6} E_n \langle \hat{\mathcal{P}}_{nn}(t)\rangle - \sum_{n=1}^{6} E_n \langle \hat{\mathcal{P}}_{nn}(0)\rangle \\ &= (E_1 - E_{BS})\langle \hat{\mathcal{P}}_{11}(t)\rangle + (E_4 - E_{BS})\langle \hat{\mathcal{P}}_{44}(t)\rangle. \end{aligned} \tag{16}$$

From this expression, we see that the on-site energy change depends only on the time-dependent populations of the SS and the energy differences $E_1 - E_{BS}$ and $E_4 - E_{BS}$ (rather than the absolute values of E_1 and E_4). [Indeed, shifting all of the site energies by the same constant results in the same dynamics during the discharge phase]. When site 4 is chosen to be the exit site (i.e., $E_4 < E_{BS} < E_1$), a higher site 4 population will lead to a more negative on-site energy change. In fact, for the dynamics displayed in Figure 1b,d, $\Delta E_{\text{on-site}}$ decreases by 50.88 cm^{-1} over the course of 1 ps (as calculated by Equation (16)), which is greater than the increase in ΔE_{exch}. This explains the decrease in the OQN energy observed in Figure 1d.

4.2. Effect of Bath Temperature

We next investigate the effect of varying the bath temperature gap on the site populations and energy flow in the QB, in an effort to find parameter sets that maximize the population of the exit site and minimize the energy loss of the OQN. To simplify our exploration of the parameter space, we vary one parameter at a time while keeping the remaining parameters fixed. First, we fix the temperature of the left bath at 300 K and vary the temperature of the right bath, while keeping the OQN, SBP, and remaining bath parameters unchanged.

In Figure 2a,c, we see that increasing the right bath temperature leads to a decrease in the population of site 4 and relatively small increases in the populations of the remaining sites. As for the energy changes (see Figure 2b,d), the OQN and left bath lose energy, but the energy loss becomes smaller and remains relatively constant, respectively, with increasing right bath temperature. On the other hand, the SBP and right bath gain energy, but the energy gain remains relatively constant and becomes smaller, respectively, with increasing right-bath temperature. When we fix the temperature of the right bath at 300 K and increase the temperature of the left bath, the behaviours of the site populations are similar to those observed in the case when the temperature of the right bath is increased, except for that of site 4 which now exhibits a substantially smaller decrease (see Figure 2c). As for the energy changes (see Figure 2d), the behaviours of the SBP and OQN energies are similar to those observed in the case when the temperature of the right bath is varied. However, the bath energies are significantly different, with the energy changes in the left and right baths becoming more negative and remaining relatively constant, respectively, with increasing left-bath temperature.

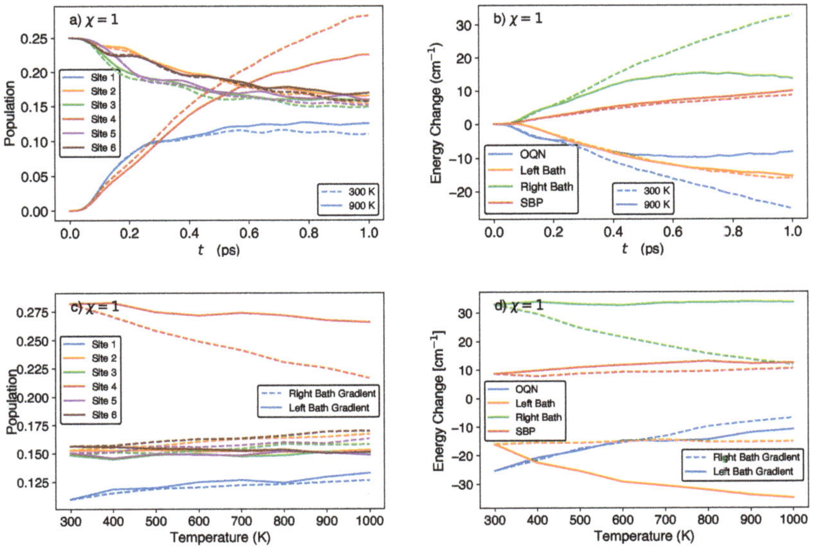

Figure 2. Site populations (**left panels**) and energy changes (**right panels**) in the discharge ($\chi = 1$) phase for different bath temperature gradients. (**a**,**b**) Time-dependent site populations and energy changes for right bath temperatures of 300 K (solid lines) and 900 K (dashed lines), with $T_L = 300$ K. (**c**,**d**) Site populations and energy changes after 1 ps for different right bath temperatures and a left bath temperature of 300 K (denoted by right bath gradient), and different left-bath temperatures and a right-bath temperature of 300 K (denoted by left bath gradient).

Increasing the bath temperature (or decreasing β) will increase the width of the initial Wigner distribution in Equation (10). This increase is particularly significant for oscillators with low frequencies. It can be verified both analytically and numerically that the ensemble averages of the initial bath kinetic energy, $\sum_j^M P_{n,j}^2$, and bath potential energy, $\sum_j^M \omega_{n,j}^2 R_{n,j}^2$, grow linearly with increasing temperature. For example, increasing the temperature of a bath from 300 K to 600 K to 900 K will increase its kinetic energy from $\approx 1.43 \times 10^4$ eV to $\approx 2.38 \times 10^4$ eV to $\approx 3.37 \times 10^4$ eV.

As seen in Figure 2d, increasing the left bath or right bath temperature will lead to similar amounts of additional energy transfer out of the bath. This temperature-driven increase in energy transfer to the OQN is relatively independent of the other parameters. As seen in Figure 3, for different site 4 energies, increasing the right bath temperature by a given amount leads to roughly the same increase in energy being transferred from the bath to the OQN. In this way, one can decrease the OQN energy loss by increasing the temperature of any bath. Finally, varying the temperature of a given bath has a larger effect on the site connected to it. For example, if we increase the left bath temperature while keeping the right bath temperature constant, there will be a larger change in the site 1 population than the site 4 population (see Figure 2). These results suggest that one could design a QB that minimizes the OQN energy loss while maintaining a relatively large site 4 population.

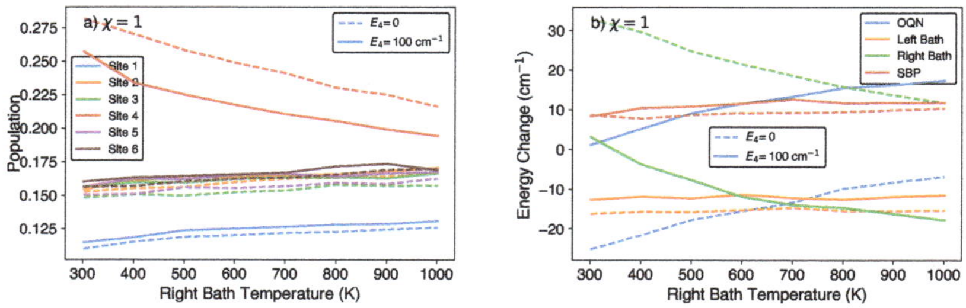

Figure 3. (**a**) Site populations and (**b**) energy changes after 1 ps in the discharge phase for different right-bath temperatures and site 4 energies, with $E_1 = 250$ cm^{-1}, $E_{i \in \{2,3,5,6\}} = 200$ cm^{-1}, and $T_L = 300$ K.

4.3. *Effect of Bath Reorganization Energy*

Next, we vary the bath reorganization energy λ_b. From Figure 4, we see that increasing the right-bath reorganization energy causes the site 1/4 population to decrease/increase, the right-bath energy to increase, and the system to lose more energy; the populations of the remaining sites, left-bath energy, and SBP energy remain relatively constant. On the other hand, we see that increasing the left-bath reorganization energy does not have a significant impact on the site population and energy changes.

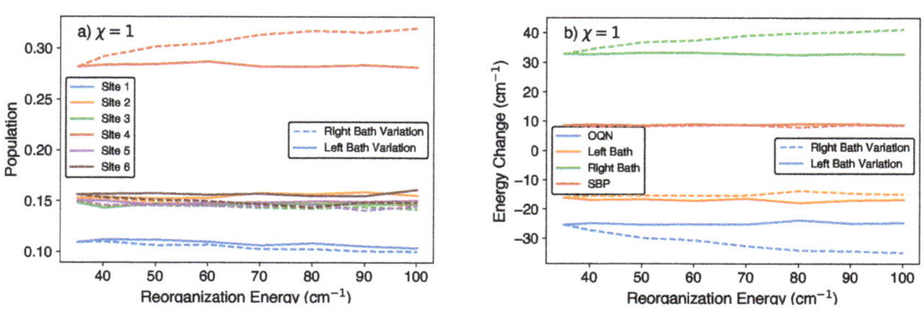

Figure 4. (**a**) Site populations and (**b**) energy changes in the discharge phase after 1 ps for different right-bath reorganization energies and a left-bath reorganization energy of 35 cm^{-1} (denoted by right-bath variation), and different left-bath reorganization energies and a right-bath reorganization energy of 35 cm^{-1} (denoted by left-bath variation).

As seen in Equation (12), increasing the bath reorganization energy λ_b will increase the coupling strength $C_{n,j}$. For example, if we increase λ_b from 35 cm^{-1} to 70 cm^{-1}, the coupling strength for each oscillator in the bath will increase by a factor of $\sqrt{2}$. Increasing the coupling strengths increases the magnitudes of the coupling terms in the equations of motion, which translates into faster energy transfer between the system and bath. As seen in Figure 4b, increasing the energy transfer rate of the right bath results in more energy transfer out of the bath after 1 ps. On the other hand, increasing the energy transfer rate of the left bath does not cause any significant changes in the left bath energy and site 1 population. This may be due to the relatively low population at site 1, viz., after 1 ps, the populations at sites 1 and 4 are $\approx$0.11 and 0.28, respectively (see Figure 1).

4.4. Effect of Site Energy

We now investigate the effects of varying the SS energies on the site populations and energy changes. Time series and results after 1 ps for different combinations of the SS energies are shown in Figure 5 and Table S1 of the SI, respectively. As seen in Figure 5a, setting the SS energies equal to each other leads to roughly equal SS populations, with the SS populations decreasing when increasing from $E_1 = E_4 = 100$ cm^{-1} to $E_1 = E_4 = 250$ cm^{-1}. When the energies of the SS are greater than those of the BSs, the BS are more populated than the SSs after 1 ps. Conversely, when the energies of the SS are lower than those of the BS, the SS are more populated than the BS after 1 ps. When the energies of the SS are equal to those of the BS, we see that the SS and BS populations approach each other over time, becoming almost equal after 1 ps. With regards to the energy changes (Figure 5b), setting the SS energies equal to each other leads to roughly equal changes in the left and right bath energies, with the bath energy changes decreasing and becoming more negative when increasing from $E_1 = E_4 = 100$ cm^{-1} to $E_1 = E_4 = 250$ cm^{-1}. More specifically, there are energy gains in the baths for $E_1 = E_4 = 100$ cm^{-1} and energy losses for the larger SS energies, with the loss increasing with increasing SS energy. Conversely, there is an energy loss from the system when $E_1 = E_4 = 100$ cm^{-1} and energy gains for the larger SS energies, with the gain increasing with increasing SS energy. The SBP energy remains mostly unchanged for the different SS energies.

In Figure 5c,d, we plot the results for $E_1 = 250$ cm^{-1} and different values of E_4. As can be seen, increasing the energy of site 4 from 0 to 100 cm^{-1} does not cause a significant change in the site 4 population; however, increasing the energy of site 4 from 100 to 250 cm^{-1} causes a $\approx$50% drop in the site 4 population after 1 ps. Moreover, when the energy of site 4 is smaller than those of the remaining sites, site 4 becomes the most populated site after 1 ps. The populations of the remaining sites do not change significantly after going from $E_4 = 0$ to 100 cm^{-1}, but they each increase by several percent after 1 ps after going from $E_4 = 100$ to 250 cm^{-1}. As for the energy changes, increasing the energy of site 4 leads to more energy transfer from the right bath to the system, which in turn causes the OQN energy to change from decreasing to increasing. When both E_1 and E_4 are greater than E_{BS} and the SS become populated, $\Delta E_{\text{on-site}}$ also becomes positive. Thus, when ΔE_{exch} is also positive, the OQN will gain energy from the baths.

Based on the results above, a few general observations may be made. First, higher site energies are associated with lower site populations. In addition, for a higher site energy, more energy is transferred from the bath to the OQN. As seen in Figure 5, for the highest E_4, the (positive) OQN energy change increases and the (negative) bath energy changes decrease, while the magnitude of the SBP energy remains low and relatively constant, i.e., energy transfer from the bath to the OQN. If one changes the site energy of a particular site, then the population of that site will be mainly affected. If two sites have the same site energy, we expect them to eventually have equal populations. Finally, when all sites have the same site energy, all sites will have the same population after a sufficiently long period of time, despite starting with different initial populations (see Figure 5a).

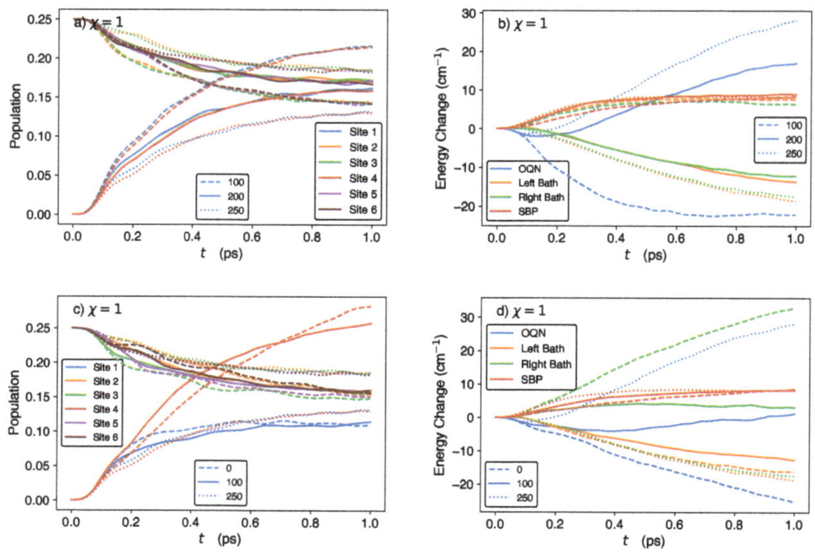

Figure 5. Time-dependent site populations (**left panels**) and energy changes (**right panels**) in the discharge phase ($\chi = 1$) for different SS energies. (**a**,**b**) Results for $E_1 = E_4 = 100$ cm^{-1}, $E_1 = E_4 = 200$ cm^{-1}, and $E_1 = E_4 = 250$ cm^{-1}. (**c**,**d**) Results for $E_1 = 250$ cm^{-1}, and $E_4 = 0$, 100, and 250 cm^{-1}. The values of the remaining parameters are $E_{i\in\{2,3,5,6\}} = 200$ cm^{-1}, $T_A = T_B = 300$ K, and $\lambda_b = 35$ cm^{-1}.

5. Concluding Remarks

In this paper, we studied the population and energy transfer dynamics of an open quantum battery model, originally proposed in Ref. [18], over a wide range of parameter regimes. In the battery's storage phase, we demonstrated that, in addition to no population leakage, there is no energy leakage from the battery into the attached baths. During the discharge phase, the changes in the populations and OQN energy are influenced by the bath temperatures, bath reorganization energies, and site energies. When increasing the temperature of one bath (while keeping the temperature of the other bath constant), we observed an increase in the energy transferred from that bath to the OQN. We found that the right bath (i.e., the bath connected to the exit site) exerts a larger influence on the exit site population than that exerted by the left bath (i.e., the bath connected to site 1) on the site 1 population. Moreover, when increasing the reorganization energy of the right bath, we observed an increase in the exit site's population and a decrease in OQN energy. On the other hand, varying the reorganization energy of the left bath does not have pronounced effect on the population and energy. Regarding the site energies, when the energy of the exit site is lower than those of the BS, the OQN energy decreases in most of the parameter regimes studied. Lowering a given site energy causes the corresponding site population to increase. When the site energies are equal, the site populations reach roughly equal values after 1 ps, despite the different initial populations.

The results of our parameter space exploration show that different parameter sets render the QB conducive to different applications. In practice, this would amount to designing the QB in such a way that its properties are consistent with those of the desired parameter set. For example, for an energy battery, one may desire that the QB gains energy from its environment during the discharge phase. As we have shown, if one sets the site energies of the SS to be larger than those of the BS, then the QB gains energy as the SS populations grow in time. On the other hand, for an excitonic battery, one may desire to maximize the population of the exit site, regardless of the change in the OQN energy.

In such a case, one could lower the exit site energy or lower the right-bath temperature while maintaining the left-bath temperature at its original value. If one would like to simultaneously reduce the loss in OQN energy and increase the exit site population, one could either increase the exit site energy or increase the left bath temperature. Overall, our findings shed light on design principles that could be used to construct different types of QBs operating between two thermal reservoirs.

Supplementary Materials: The following supporting information can be downloaded at: https://www.mdpi.com/article/10.3390/molecules29040889/s1. Total energy drift, effect of varying surface site energies, time derivative of the bath and network-bath energies, and time-dependent quantum coherences.

Author Contributions: Conceptualization, Z.L. and G.H.; methodology, Z.L. and G.H.; software, Z.L.; validation, Z.L. and G.H.; formal analysis, Z.L.; investigation, Z.L. and G.H.; resources, G.H.; writing—original draft preparation, Z.L.; writing—review and editing, Z.L. and G.H.; visualization, Z.L.; supervision, G.H.; funding acquisition, G.H. All authors have read and agreed to the published version of the manuscript.

Funding: This research was funded by the Natural Sciences and Engineering Research Council of Canada (grant number RGPIN-2020-05977).

Institutional Review Board Statement: Not applicable.

Data Availability Statement: Data are contained within the article and supplementary materials.

Acknowledgments: The authors would like to thank Mahmoud Mahdian for useful discussions during the early stages of this work.

Conflicts of Interest: The authors declare no conflicts of interest. The funders had no role in the design of the study; in the collection, analyses, or interpretation of data; in the writing of the manuscript; or in the decision to publish the results.

References

1. Stefanatos, D. Optimal efficiency of a noisy quantum heat engine. *Phys. Rev. E* **2014**, *90*, 012119. [CrossRef]
2. Binder, F.C.; Vinjanampathy, S.; Modi, K.; Goold, J. Quantacell: Powerful charging of quantum batteries. *New J. Phys.* **2015**, *17*, 075015. [CrossRef]
3. Campaioli, F.; Pollock, F.A.; Binder, F.C.; Céleri, L.; Goold, J.; Vinjanampathy, S.; Modi, K. Enhancing the Charging Power of Quantum Batteries. *Phys. Rev. Lett.* **2017**, *118*, 150601. [CrossRef]
4. Ferraro, D.; Campisi, M.; Andolina, G.M.; Pellegrini, V.; Polini, M. High-Power Collective Charging of a Solid-State Quantum Battery. *Phys. Rev. Lett.* **2018**, *120*, 117702. [CrossRef]
5. Barra, F. Dissipative Charging of a Quantum Battery. *Phys. Rev. Lett.* **2019**, *122*, 210601. [CrossRef]
6. Farina, D.; Andolina, G.M.; Mari, A.; Polini, M.; Giovannetti, V. Charger-mediated energy transfer for quantum batteries: An open-system approach. *Phys. Rev. B* **2019**, *99*, 035421. [CrossRef]
7. Rossini, D.; Andolina, G.M.; Polini, M. Many-body localized quantum batteries. *Phys. Rev. B* **2019**, *100*, 115142. [CrossRef]
8. Caravelli, F.; Yan, B.; Garcia-Pintos, L.P.; Hamma, A. Energy storage and coherence in closed and open quantum batteries. *Quantum* **2021**, *5*, 505. [CrossRef]
9. Mazzoncini, F.; Cavina, V.; Andolina, G.M.; Erdman, P.A.; Giovannetti, V. Optimal control methods for quantum batteries. *Phys. Rev. A* **2023**, *107*, 032218. [CrossRef]
10. Alicki, R.; Fannes, M. Entanglement boost for extractable work from ensembles of quantum batteries. *Phys. Rev. E* **2013**, *87*, 042123. [CrossRef] [PubMed]
11. Giorgi, G.L.; Campbell, S. Correlation approach to work extraction from finite quantum systems. *J. Phys. B* **2015**, *48*, 035501. [CrossRef]
12. Andolina, G.M.; Keck, M.; Mari, A.; Campisi, M.; Giovannetti, V.; Polini, M. Extractable Work, the Role of Correlations, and Asymptotic Freedom in Quantum Batteries. *Phys. Rev. Lett.* **2019**, *122*, 047702. [CrossRef]
13. Shi, H.L.; Ding, S.; Wan, Q.K.; Wang, X.H.; Yang, W.L. Entanglement, Coherence, and Extractable Work in Quantum Batteries. *Phys. Rev. Lett.* **2022**, *129*, 130602. [CrossRef] [PubMed]
14. Dou, F.Q.; Lu, Y.Q.; Wang, Y.J.; Sun, J.A. Extended Dicke quantum battery with interatomic interactions and driving field. *Phys. Rev. B* **2022**, *105*, 115405. [CrossRef]
15. Breuer, H.P.; Petruccione, F. *The Theory of Open Quantum Systems*; Oxford University Press: Oxford, UK, 2007. [CrossRef]
16. Weiss, U. *Quantum Dissipative Systems*; World Scientific: Singapore, 2012; Volume 13.
17. De Vega, I.; Alonso, D. Dynamics of non-Markovian open quantum systems. *Rev. Mod. Phys.* **2017**, *89*, 015001. [CrossRef]

18. Liu, J.; Segal, D.; Hanna, G. Loss-free excitonic quantum battery. *J. Phys. Chem. C* **2019**, *123*, 18303–18314. [CrossRef]
19. Lidar, D.A. Review of Decoherence-Free Subspaces, Noiseless Subsystems, and Dynamical Decoupling. In *Quantum Information and Computation for Chemistry*; John Wiley & Sons, Ltd.: Hoboken, NJ, USA, 2014; pp. 295–354. [CrossRef]
20. Jamiołkowski, A.; Kamizawa, T.; Pastuszak, G. On invariant subspace in quantum control systems and some concepts of integrable quantum systems. *Int. J. Theor. Phys.* **2015**, *54*, 2662–2674. [CrossRef]
21. Buča, B.; Prosen, T. A note on symmetry reductions of the Lindblad equation: Transport in constrained open spin chains. *New J. Phys.* **2012**, *14*, 073007. [CrossRef]
22. Thingna, J.; Manzano, D.; Cao, J. Dynamical signatures of molecular symmetries in nonequilibrium quantum transport. *Sci. Rep.* **2016**, *6*, 1–11. [CrossRef]
23. Albert, V.V.; Jiang, L. Symmetries and conserved quantities in Lindblad master equations. *Phys. Rev. A* **2014**, *89*, 022118. [CrossRef]
24. Suter, D.; Álvarez, G.A. Colloquium: Protecting quantum information against environmental noise. *Rev. Mod. Phys.* **2016**, *88*, 041001. [CrossRef]
25. Zanardi, P.; Rasetti, M. Noiseless quantum codes. *Phys. Rev. Lett.* **1997**, *79*, 3306. [CrossRef]
26. Lidar, D.A.; Chuang, I.L.; Whaley, K.B. Decoherence-free subspaces for quantum computation. *Phys. Rev. Lett.* **1998**, *81*, 2594. [CrossRef]
27. Lidar, D.A.; Birgitta Whaley, K. Decoherence-Free Subspaces and Subsystems. In *Irreversible Quantum Dynamics*; Benatti, F., Floreanini, R., Eds.; Springer: Berlin/Heidelberg, Germany, 2003; pp. 83–120. [CrossRef]
28. Beige, A.; Braun, D.; Tregenna, B.; Knight, P.L. Quantum computing using dissipation to remain in a decoherence-free subspace. *Phys. Rev. Lett.* **2000**, *85*, 1762. [CrossRef] [PubMed]
29. Xue, P.; Xiao, Y.F. Universal quantum computation in decoherence-free subspace with neutral atoms. *Phys. Rev. Lett.* **2006**, *97*, 140501. [CrossRef]
30. Manzano, D.; Hurtado, P.I. Symmetry and the thermodynamics of currents in open quantum systems. *Phys. Rev. B* **2014**, *90*, 125138. [CrossRef]
31. Manzano, G.; Giorgi, G.L.; Fazio, R.; Zambrini, R. Boosting the performance of small autonomous refrigerators via common environmental effects. *New J. Phys.* **2019**, *21*, 123026. [CrossRef]
32. Gherardini, S.; Campaioli, F.; Caruso, F.; Binder, F.C. Stabilizing open quantum batteries by sequential measurements. *Phys. Rev. Res.* **2020**, *2*, 013095. [CrossRef]
33. Tejero, Á.; Thingna, J.; Manzano, D. Comment on "Loss-Free Excitonic Quantum Battery". *J. Phys. Chem. C* **2021**, *125*, 7518–7520. [CrossRef]
34. Manzano, D.; Hurtado, P. Harnessing symmetry to control quantum transport. *Adv. Phys.* **2018**, *67*, 1–67. [CrossRef]
35. Liu, J.; Hanna, G. Efficient and deterministic propagation of mixed quantum-classical Liouville dynamics. *J. Phys. Chem. Lett.* **2018**, *9*, 3928–3933. [CrossRef] [PubMed]
36. Liu, Z.; Sergi, A.; Hanna, G. DECIDE: A Deterministic Mixed Quantum-Classical Dynamics Approach. *Appl. Sci.* **2022**, *12*, 7022. [CrossRef]
37. Liu, J.; Hsieh, C.Y.; Segal, D.; Hanna, G. Heat transfer statistics in mixed quantum-classical systems. *J. Chem. Phys.* **2018**, *149*, 224104. [CrossRef]
38. Liu, J.; Segal, D.; Hanna, G. Hybrid quantum-classical simulation of quantum speed limits in open quantum systems. *J. Phys. A* **2019**, *52*, 215301. [CrossRef]
39. Carpio-Martínez, P.; Hanna, G. Nonequilibrium heat transport in a molecular junction: A mixed quantum-classical approach. *J. Chem. Phys.* **2019**, *151*, 074112. [CrossRef]
40. Carpio-Martínez, P.; Hanna, G. Quantum bath effects on nonequilibrium heat transport in model molecular junctions. *J. Chem. Phys.* **2021**, *154*, 094108. [CrossRef]
41. Wigner, E. On the Quantum Correction For Thermodynamic Equilibrium. *Phys. Rev.* **1932**, *40*, 749–759. [CrossRef]
42. Imre, K.; Özizmir, E.; Rosenbaum, M.; Zweifel, P. Wigner method in quantum statistical mechanics. *J. Math. Phys.* **1967**, *8*, 1097–1108. [CrossRef]
43. Wang, H.; Thoss, M.; Miller, W.H. Systematic convergence in the dynamical hybrid approach for complex systems: A numerically exact methodology. *J. Chem. Phys.* **2001**, *115*, 2979–2990. [CrossRef]
44. Huo, P.; Coker, D.F. Semi-classical path integral non-adiabatic dynamics: A partial linearized classical mapping Hamiltonian approach. *Mol. Phys.* **2012**, *110*, 1035–1052. [CrossRef]
45. Ishizaki, A.; Fleming, G.R. Theoretical examination of quantum coherence in a photosynthetic system at physiological temperature. *Proc. Natl. Acad. Sci. USA* **2009**, *106*, 17255–17260. [CrossRef] [PubMed]
46. Shanks, E.B. *Higher Order Approximations of Runge-Kutta Type*; Technical Report NASA-TN-D-2920; NASA Marshall Space Flight Center NTRS Document ID: 19650022581; NTRS Research Center: Moffett Field, CA, USA, 1965.

molecules

MDPI

Article

Exploring Spin Distribution and Electronic Properties in FeN_4-Graphene Catalysts with Edge Terminations

Ismail Can Oguz, Frederic Jaouen and Tzonka Mineva *

ICGM, Univ. Montpellier, 34293 Montpellier, France; i.c.oguz@differ.nl (I.C.O.);
frederic.jaouen@umontpellier.fr (F.J.)
* Correspondence: tzonka.mineva@enscm.fr

Abstract: Understanding the spin distribution in FeN_4-doped graphene nanoribbons with zigzag and armchair terminations is crucial for tuning the electronic properties of graphene-supported non-platinum catalysts. Since the spin-polarized carbon and iron electronic states may act together to change the electronic properties of the doped graphene, we provide in this work a systematic evaluation using a periodic density-functional theory-based method of the variation of spin-moment distribution and electronic properties with the position and orientation of the FeN_4 defects, and the edge terminations of the graphene nanoribbons. Antiferromagnetic and ferromagnetic spin ordering of the zigzag edges were considered. We reveal that the electronic structures in both zigzag and armchair geometries are very sensitive to the location of FeN_4 defects, changing from semi-conducting (in-plane defect location) to half-metallic (at-edge defect location). The introduction of FeN_4 defects at edge positions cancels the known dependence of the magnetic and electronic proper-ties of undoped graphene nanoribbons on their edge geometries. The implications of the reported results for catalysis are also discussed in view of the presented electronic and magnetic properties.

Keywords: FeN_4-C catalysts; spin; band structure; zigzag graphene nanoribbons; armchair graphene nanoribbons; DFT

1. Introduction

The past two decades have seen significant attention directed toward controlling electronic spin moments in graphene nanostructures. It began with the theoretical prediction of spin-polarizable mono-hydrogenated edge states in zigzag graphene nanoribbons (ZGNRs) [1], paving the way for creating magnetic graphene finite-size structures. These structures hold promise for applications in spintronics and quantum information technologies. However, synthesizing magnetic monohydrated ZGNRs remains challenging [2,3], leading to extensive theoretical [3–11] and experimental efforts [12–18]. The objective is to understand, modify, and control the spin-density distributions in nanostructured graphene. Among the strategies for shaping graphene's magnetic properties, doping with heteroatoms has emerged as a general method for tuning magnetic, electronic, optical, and transport properties [9,18–23]. Computational studies using density functional theory (DFT) revealed that doping GNRs with nitrogen, boron, or transition metal (TM) atoms at specific positions can induce thermodynamically stable ferromagnetic GNRs with spin-polarized edge carbons, depending on the edge geometry (zigzag or armchair), ribbon widths, location, and density of dopant atoms [9,20,21,24–27]. These studies also agree that GNRs with dopants at the edges are generally more energetically stable than those doped far from the edge (in-plane) of the graphene nanoribbons.

The ability of graphene to stabilize various dopants is also used in catalysis by employing graphene to host a variety of active sites, including non-metal dopants and a broad range of single metal atoms, from 3d transition metals to heavier elements, such as Sn and platinum-group metals [28–30]. In particular, co-doped graphene with transition metals and nitrogen has been emerging as a promising electrocatalyst for oxygen reduction reactions (ORR) [22,31–33], CO_2 conversion [34–36], or nitrate reductions [36]. TMNx-graphene has been identified as an efficient catalyst because of its spin polarization [36,37]. In particular, FeN_4 moieties that are covalently integrated in carbonaceous matrices have been established as the most promising precious-metal-free active sites in proton exchange membrane fuel cell cathodes for ORR [22,32,38–40]. Unrevealing the origins behind the good reactivity of Fe-N-C catalysts has become an important milestone for a more rational design of non-precious metal catalysts. The FeN_xC_y coordination structures and iron, magnetic, and electronic properties are among the key factors that determine the reactivity of Fe-N-C catalysts. Experimental techniques, such as extended X-ray absorption fine structure (EXAFS), X-ray absorption near-edge structure (XANES) spectroscopy, X-ray photoelectron spectroscopy (XPS), and Mössbauer spectroscopy, are commonly used to discern the geometrical, electronic, and magnetic properties of synthesized Fe-N-C catalysts. Meanwhile, DFT calculations applied to truncated or extended FeN_4-graphene models serve to provide insights into the local iron environment and its electronic structure [32,41–46].

The pyridinic (FeN_4C_{10}) and porphyrinic (FeN_4C_{12}) configurations are widely recognized as the predominant coordination geometries of the active sites in pyrolyzed Fe-N-C catalysts for the oxygen reduction reaction (ORR). By analyzing the computed and measured Mössbauer quadrupole splitting data, the presence of high-spin Fe(III)-N_4C_{12} porphyrinic structures and low- or medium-spin Fe(II)-N_4C_{10} pyridinic structures in pyrolyzed Fe-N-C catalysts has been established [42]. DFT studies have revealed the thermodynamic stability of spin-polarized at-edge TM-C_xN_y defects [21,22,31,37,47]. A high magnetic moment computed in FeN_4-AGNR and FeN_4-ZGNR was attributed to parallel spin moments on the localized 3d electrons of Fe atom and 2p electrons of C atoms [47]. More recent theoretical studies have correlated the ORR activity with the iron magnetic moments in FeN_4-graphene, owing to the hybridization between spin-polarized Fe3d and O2p orbitals in the activated OH complex [22]. The enhancement in the performance of the FeN_4-graphene catalyst was attributed to an alteration in FeN_4-graphene band structure from metallic to half-metallic [48], meaning that the spin-up bands have isolating character, whereas spin-down bands have conducting character, or vice versa. In light of these studies, the spin distribution in FeN_4-graphene structures appears to play a pivotal role in their electrocatalytic activity; however, the concomitant spin polarization of graphene edge carbons, possibly occurring in truncated FeN_4-graphene models, has been considered in very few cases [47]. Comprehending the spin interactions between co-doped iron–nitrogen states and carbon edge states remains an open question, yet it is crucial for regulating the electronic properties and spin density at metal active sites within Fe-N-C materials, particularly for their application as electrocatalysts.

In this study, we examine the spin distribution between iron and carbon atoms, and the sub-band profile of both Zigzag Graphene Nanoribbons and Armchair Graphene Nanoribbons. These ribbons integrate pyridinic-like FeN_4 defects positioned at various locations between the ribbon's center and one of its edges. This study explores both ferromagnetic and antiferromagnetic spin ordering at the edges concerning the FeN_4-GNR's thermodynamic stability, and electronic and magnetic properties.

2. Results

2.1. Formation and Stability of Fe(II)N_4 in ZGNR and AGNR Models

Initially, we concentrated on the stability of the pyridinic Fe(II)N_4 active site, ranging from bulk sites to edge configurations in the graphene nanoribbons with zigzag and armchair terminations, shown in Figure 1, on the left and right sides, respectively. The dangling bonds of the edge carbon atoms are saturated with hydrogen (H).

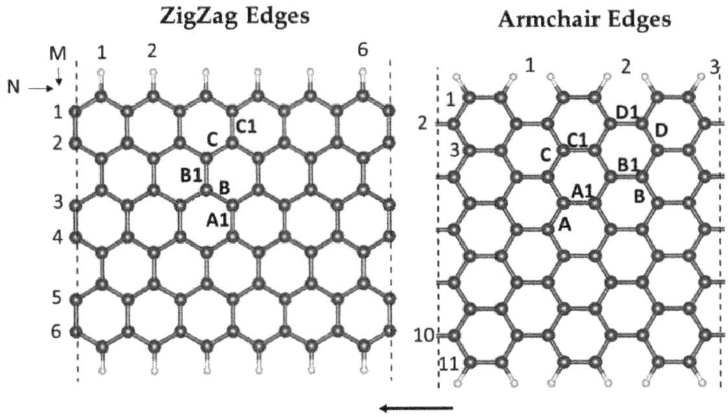

Figure 1. Atomic structure of zigzag (**left**) and armchair (**right**) graphene nanoribbons. The labelled carbon bonds indicate the positions of the vacancies, created by removing two bound carbons. FeN_4 active sites are subsequently integrated into the double vacancies (see text). The arrows indicate the periodicity direction of ZGNRs and AGNRs. The N and M indicate, respectively, the number of edge atoms and the width of GNR. The atom color code is the following: brown for carbon and white for hydrogen.

Conventionally, ZGNRs are described by the number of atoms along a row and the width of the model, denoted by the notation (N × M). Here, 'N' represents the number of atoms on the edge row, and 'M' indicates the width of the model. The notation we employed was first introduced by Cervantes-Sodi et al. [49]. According to this notation, the ZGNR model shown in Figure 1 was defined as a (6 × 6) ZGNR, and the AGNR model was designated as an (3 × 11) AGNR. To prevent interactions between an Fe atom and its periodic image, graphene nanoribbon models were selected with sufficient size: the distance between the Fe atoms and their periodic images in the adjacent cell was maintained at 14.83 Å for zigzag and 12.83 Å for armchair configurations.

The FeN_4-GNR unit cells were constructed after the formation of a double vacancy (DV), by subtracting a pair of C atoms and replacing the four dangling C atoms with nitrogens. The Fe atom was then embedded in the DV center, coordinated to four N atoms. We explored five FeN_4-ZGNR and eight FeN_4-AGNR positions. The DVs were chosen to represent two different orientations of FeN_4 with respect to the periodicity axis: orientations perpendicular or aligned with the periodicity axis are labelled as A1, B1, C1, and D1 in Figure 1, hereafter called "direct"; and orientations tilted at 30° or 60° to the ZGNR and AGNR periodicity axes are labelled as A, B, C, and D in Figure 1, hereafter called "tilted". The energetic and magnetic properties of bare or single-atom-doped GNRs may vary with the ribbon width. Strong dependence of dopant binding and formation energies were reported for very thin ribbons with width M = 2 carbons in the row, but going beyond this limit, the effect of the ribbon widths decreased rapidly. The width of the models considered here (M = 6 for ZGNR and M = 11 for AGNR) is therefore sufficient to exclude a significant variation of the computed properties with variations of the ribbon widths.

The reported formation energies (E_f) and binding energies (E_b) displayed in Figure 2 confirm the thermodynamically stable integration of FeN_4 defects at all thirteen examined positions within both GNRs. This aligns with the previously established good thermodynamic stability of covalently integrated FeN_4 moieties in different models of graphene nanostructures [17–19,22,44]. Note that, in our comparative study, E_f and E_b were computed for the lowest energy structures, whose edge spin configurations were with a ferromagnetic (FM) order. The translocation of the Fe center towards the edge results in the suppression of spin on one GNR side, thereby rendering the antiferromagnetic (AFM) order at the edge untenable. Consequently, to maintain a uniform standard across all models, configurations derived from ferromagnetic ordering at the edges were used to ascertain energy stability. The A1 configuration is an exception because the antiferromagnetic spin orientations of the edges were found to be more stable by 23 meV than the ferromagnetic spin ordering. The definitions used to compute E_f and E_b are reported in the Methods section.

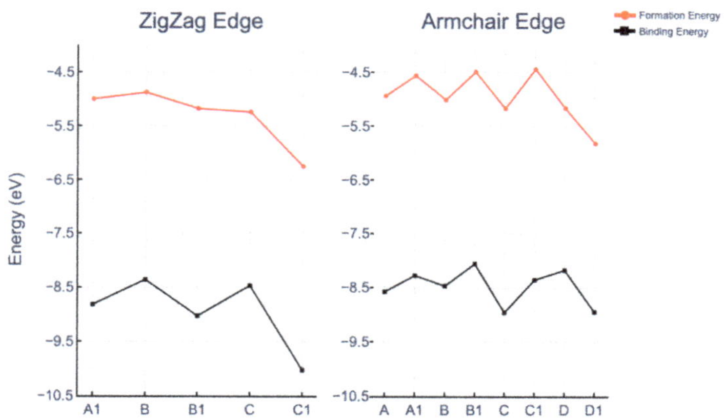

Figure 2. Formation and binding energies of FeN_4 moieties in ZGNR (**left**) and AGNR (**right**).

The formation energies associated with the FeN_4 defect varied depending on its orientation. In the case of ZGNRs, the "tilted" orientations (points B and C in Figure 2, left-hand side) were revealed to be less stable than the defects with "direct" orientations (points B1 and C1 in Figure 2, left-hand side). For the armchair configurations, "tilted" orientations generally exhibited greater stability than their "direct" counterparts (compare A, B, and C with A1, B1, and C1 in Figure 2, right-hand side). The only exception to this trend was the stronger stability of the "direct" FeN_4 in D1 position (at the edge) in comparison with the "tilted" defect in position D (D1 and D points in Figure 2, right-hand side). This can be associated with the different FeN_4 environments in the D and D1 locations, which apparently contributed more strongly to the FeN_4 stabilization than the specific orientation of FeN_4. In the D-placed defect with "tilted" orientation, one nitrogen became pyrrolic-like (at the edge) and the other three nitrogens remained pyridinic, whereas in the D1-placed defect, two nitrogens were pyrrolic-like. The D1 defect had the same carbon environment as the most stable C1 defect in ZGNR, with two pyridinic and two pyrrolic-like nitrogens. Thus, increasing the number of pyrrolic-like nitrogen atoms at the edges increased the thermodynamic stability of FeN_4-GNRs. This aligns with Li et al.'s findings [27], where nitrogen defects near the edges of ZGNRs were energetically favorable, and pyrrole-like defects possessed even lower formation energies than pyridinic-like defects. Our findings on Fe-centered defect stabilization in graphene nanoribbons are consistent with Holby et al.'s observation [28] of preferential edge stabilization of Fe-pyridinic vacancy complexes (FeN_3) in GNRs.

Overall, a relatively small variation (within 15%) in the formation and binding energies of FeN_4 from in-plane positions A and A1 to the edge positions C1 and D1 was obtained. Fluctuations in formation energy were a recurrent theme in the study of nitrogen or metal-substituted graphene nanoribbons [27,28,50,51], underscoring the edge configuration as the most stable site. These investigations have also considered pyridine-like structures, such as 3NV and 4N defects, as in our work, which also demonstrates the highest stability at the edges of zigzag graphene nanoribbons.

2.2. Ferromagnetic and Antiferromagnetic Spin Ordering at the ZGNR Edges

The ferromagnetic (FM) and antiferromagnetic (AFM) spin orderings were considered for the minimum energy ZGNRs. For comparison, the magnetic and electronic features of the pristine ZGNR and AGNR were computed at the same theoretical basis. It was generally reported from the DFT calculations of ZGNRs that the FM ordering of the spin-polarized orbitals of the edge carbons was slightly less stable than the AFM phase. Our results are in line with this finding, reporting AFM ordering as more stable by 59 meV than the FM spin orientation. The average spin moment localized at every carbon atom at the zigzag edges was 0.23 μ_B (Figure S1a). The FM band structure in Figure S1b shows a metallic character because of the crossing of spin-up and spin-down states at the Fermi level. In this band diagram, the spin-up states are shifted to the conduction band and the spin-down states are dominating in the valence band zone. Consistent with earlier findings [1,3,4,7,9,10], the AFM structure (refer to Figure S1c) demonstrated a semiconductor-like nature, displaying a band gap of 0.54 eV with degenerate spin-up and spin-down channels that overlapped. At each edge carbon, the magnetic moment was ± 0.24 μ_B, resembling closely those in the FM ZGNR. Analysis of band-decomposed charge density plots in Figure S2 revealed that the bands near the Fermi level were primarily composed of spin-polarized p_z orbitals originating from the edge carbons.

Fe(II)N_4 defects located at the central DV (position A1 in Figure 1) in ZGNR decreased negligibly the spin moments at the edge carbons with respect to the pristine GNRs, as revealed by the results in Figure 3, left- and right-hand sides for the FM and AFM ordering, respectively. The absolute value of the spin moment at the iron atom was 2.04μ_B, for both the AFM and FM ordering of the carbon atoms at the edges (Figure 3). However, the overall AFM-FeN_4-ZGNR structure was energetically more stable by 23 meV compared with FM-FeN_4--ZGNR. The total magnetic moment of FeN_4-ZGNRs therefore resulted from the Fe3d and C p_z spin-polarized states and became 4.54 and 2.0μ_B, respectively, for FM and AFM ordering of the edges.

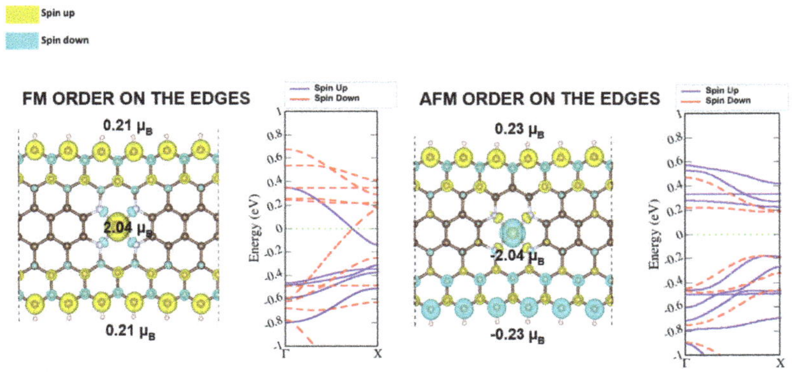

Figure 3. Spin density and band diagram of FM- (**left**) and AFM-ordered (**right**) carbon edges in FeN4-ZGNRs. Yellow and turquoise colors represent the spin-up and spin-down distribution over lattice structure. Blue and dashed red colors represent the spin-up and spin-down states in band diagram.

Despite the FeN_4 integration into graphene, the band structures preserved their character as in the pristine ZGNR (Figure S1). The FM edges determined a conducting band (Figure 3, left-hand side) and the AFM edges determined a semi-conducting band structure (Figure 3, right-hand side). Doping with FeN_4 at the in-plane position in ZGNRs did not affect the electronic structure and the magnetic distribution on edge carbons, but only the total magnetic moment of the ribbon.

2.3. Magnetic and Electronic Properties as a Function of FeN_4 Location and Edge Termination

The evolution of the spin-density and sub-band structures of FeN_4 with different orientations and positions was explored considering the more stable AFM-ZGNR, with opposite spin-moment orientations at the edges and AGNR with non-magnetic (with zero spin moments) edges.

The computed spin-density distributions and band diagrams for the five positions of the FeN_4 in ZGNRs are presented in Figure 4a,b and those of the four "direct" positions FeN_4-AGNRs are presented in Figure 4c,d. The band-gap values presented in Figure 4b,d were calculated using the GGA-PBE functional, known to underestimate band gaps for systems with strong electronic correlations. However, while PBE may not yield precise absolute band-gap values, it reliably captures trends in the band structures variations with FeN_4 location and orientation.

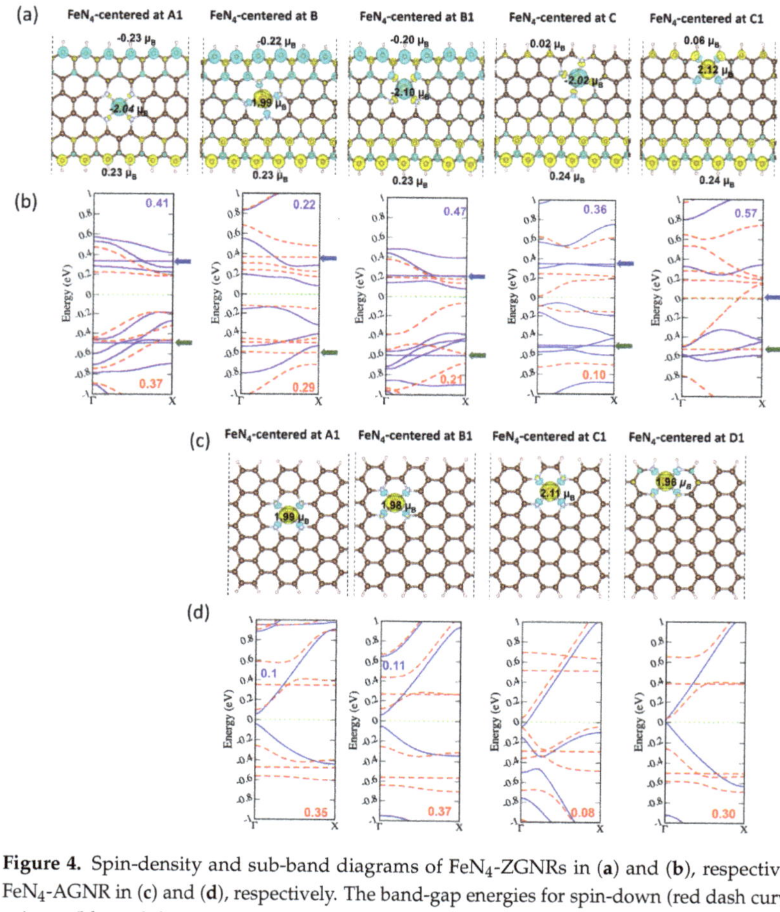

Figure 4. Spin-density and sub-band diagrams of FeN_4-ZGNRs in (**a**) and (**b**), respectively, and FeN_4-AGNR in (**c**) and (**d**), respectively. The band-gap energies for spin-down (red dash curves) and spin-up (blue solid curves) channels are in eV. In (**b**), the blue arrows point to the d_{z^2} states and the green arrows point to the $d_{x^2-y^2}$ ("direct" orientation) or d_{xy} ("tilted" orientation) states. The Fermi energy is set to zero. Yellow and turquoise colors represent the spin-up and spin-down distribution over lattice structure.

The analysis focused first on zigzag edge models. To prevent miscalculation, adjustments were made to the unit cell structure, specifically due to the FeN_4 moiety in graphene, considering potential strain-induced modifications in spin moments and edge direction that can influence the transition from antiferromagnetic (AFM) to ferromagnetic (FM) ordering. To mitigate strain effects, a two-step approach was employed. Initially, geometric cells of zigzag models were relaxed for unit cell adjustment, seeking an optimal x-axis cell parameter. The final calculation used a fixed cell with the determined unit cell parameter. Notably, "direct" FeN_4 defects expanded the unit cell by approximately 0.1 Å up to 14.89 Å, while "tilted" FeN_4 moieties contracted the unit cell by approximately 0.08 Å down to 14.71 Å, with respect to the undoped zigzag nanoribbon model (Figure 1), with an initial unit cell parameter of 14.79 Å. This methodology was not applied to AGNR models, as there is no effect of strain on spin moments on the non-magnetic armchair models. Discussing first the FeN_4-ZGNR structures, it is worth noting that the spin distribution on the carbon edges was not affected by the in-plane defects in the baseline A1, B, and B1 locations. The total magnetic moment of these structures was equal to the magnetic moment at the Fe 3d electrons, because of the opposite spin directions of edge carbons that cancel when integrating the spin density over all atoms in the unit cell. The situation changed for the near-edge (C1) and on-edge (C) FeN_4 locations, for which the carbon spin moments were quenched. Even more, not only did the spin density at each carbon drop down, but the sign of the spin moment also changed from $-0.20\ \mu_B$ (B1 in Figure 4a) to $0.02\ \mu_B$ (from B1 to C in Figure 4a). Consequently, the near-edge FeN_4-ZGNR entire structure became non-magnetic because the Fe spin-down moment cancelled the spin-up moment of the carbons at the opposite edge. On the contrary, the magnetic moment of the on-edge FeN_4-ZGNR structure (FeN_4 centered at C in Figure 4a) increased up to 4 μ_B (Table 1), resulting mainly from the parallel spin moments on the iron and on the opposite edge carbons. Interestingly, the spin quenching of the carbons at the edge close to the FeN_4 site did not affect the absolute value of the Fe spin moment. Previously studied N-doped ZGNRs edges were established as magnetic [26], different from our findings for FeN_4-doped ZGNR edges. Other DFT studies have reported on the spin suppression of edge carbon sites in Ti-, Pt-, and Au-doped ZGNR edges [52–54].

Table 1. Total magnetic moment (μ_{Total}) and Fe magnetic moment (μ_{Fe}) in μ_B of FeN_4-ZGNRs with initial AFM spin ordering at the edges.

FeN_4 Position	A1	B	B1	C	C1
μ_{Total}	−2.00	2.02	−1.97	−0.09	4.00
μ_{Fe}	−2.04	1.99	−2.10	−2.02	2.12

The band structures depicted in Figure 4b also exhibited sensitivity to the FeN_4 location, which was especially noticeable in the spin-down channel (red curves in Figure 4b). Localized flat-band characteristics due to $d_{x^2-y^2}$, d_{xy}, and d_{z^2} states were observed in the band structures in all FeN_4-ZGNRs, as shown by the blue and green arrows pointing to these d-bands in Figure 4b. The unoccupied d_{z^2} localized band (pointed by a blue arrow in Figure 4b) shifted downward the Fermi level when moving the FeN_4 defect from A1 to B1 to C1. The d_{z^2} unoccupied band localized at energy levels of 0.34, 0.21, and 0.01 eV in the A1, B1, and C1 configurations, respectively. The contributions of d_{yz} and d_{xz} in the bands were intertwined with the p_z orbital contribution from carbon. This results in hybridization between the Cp_z and Fe d_{xz}/d_{yz} orbitals, as demonstrated for the C1 FeN_4-ZGNR structure in Figure 5.

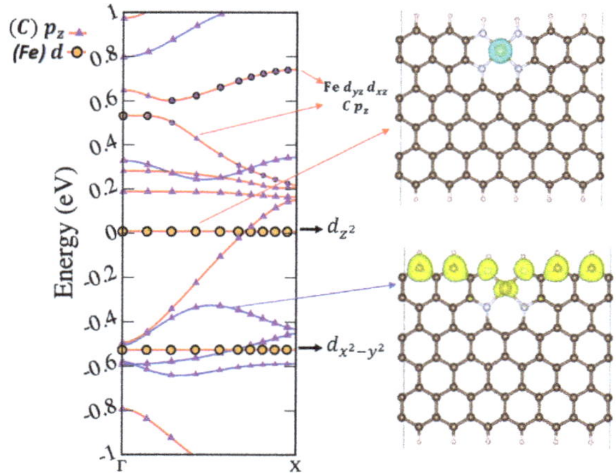

Figure 5. Charge-density decomposition of the sub-bands in the at-edge FeN_4-ZGNR, showing the fat bands for the states labelled as C p_z and Fe 3d orbitals (**left**) side. The Fermi level in the sub-band diagram is set to zero. Spin-down (turquoise color) Fe $3d_{z^2}$ charge density is plotted in the upper (**right**) panel and the spin-up (yellow) Cp_z charge-density distribution is plotted in the right down panel.

As the iron dopant was shifted towards the zigzag ribbon edge, the spin-down band gap diminished and closed entirely in the on-edge FeN_4-ZGNR configuration. Consequently, the band structure became half-metallic, displaying a metallic character along the spin-down channel and a semi-conducting character with a 0.57 eV band gap along the spin-up channel (see Figure 4b, last column). Pt-doping at the ZGNR edges was also found to promote the half-metallic state [52]. The charge-density decomposition, presented in fat band in Figure 5, revealed that the half-metallic character was due to the spin-down Fe dz^2 electrons at the Fermi level (the spin-density in the upper-right panel in Figure 5) and the Cp_z electrons, crossing the Fermi level. The valence band was formed by the occupied p_z electrons with opposite spins and the Fe $d_{x^2-y^2}$ electrons with spin down. The $Fe3d_{yz}$ and $3d_{xz}$ orbitals were in the conduction bands. Assuming the validity of the semi-empirical rule correlating the smaller distance of the center of the 3d band of transition-metal surfaces with their enhanced catalytic activity [53] holds also for the half-metallic FeN_4-doped graphene, a high reactivity could be suggested for the on-edge FeN_4-ZGNRs, due to the 0.01 eV spacing between Fermi level and the center of the 3d band (see Figure 5).

The total spin moment in the doped structures with armchair termination was carried out essentially by Fe3d electrons (see Figure 4c for "direct" oriented defects and Figure S3 for "tilted" oriented defects), regardless of the location of the FeN_4 defect. This result differs from the previously reported magnetization of AGNR edges due to doping with isolated transition metal and nitrogen atoms [21]. Nevertheless, the FeN_4 dopants did not cause spin polarization of the armchair carbons, the band structure in the "direct" FeN_4-dopants changes from semi-conducting in the in-plane-doped (position A) AGNR to half-metallic in the near- and on-edge-doped structures (see Figure 4d). For comparison, a band gap of 0.2 eV was calculated for the pristine AGNR, shown in Figure S4. The changes from semi-conducting to half-metallic states were not found for the "tilted" FeN_4-AGNRs as reported in Figure S3. It follows that the FeN_4 dopant orientation played a role on the AGNR band-structure features, but did not polarize the spin states.

The validity of the above discussed models and results was verified for a cell with a double length along the periodicity direction and integrating two FeN_4 sites, the ($2 \times 1 \times 1$) supercell model, of in-plane "direct" FeN_4-ZGNRs and "tilted" FeN_4-AGNRs. This allowed also to explore the anti-parallel orientation of the spin moments on the two Fe sites, thus the AFM doped zigzag ribbons. Except the changes in the total magnetic moment due to the presence of two FeN_4 defects in the supercell, the charge-densities and band-diagrams in Figure S5 (FeN_4-ZGNRs) and Figure S6 (FeN_4-AGNRs) show the same characteristics as those obtained for the respective unit cells, discussed in detail above. Moreover, considering two FeN_4 dopants on the opposite ZGNR edges (see Figure S7) or two FeN_4 on-edge defects at the same edge in a (2 × 1 × 1) supercell in Figure S8 always resulted in spin-quenching of the edge carbons, which is demonstrated with the nearly zero magnetic moments per carbon atom as reported in the Figure S8.

2.4. Discussion: Implication for Catalysis

Our results demonstrate that the coupling between iron magnetic moment and the magnetic moments of carbon atoms at the graphene edges significantly modifies the electronic structure of the materials depending on three factors: (i) graphene edge-geometries, i.e., zigzag or armchair; (ii) location of FeN_4 in the graphene nanoribbon; and (iii) orientation of FeN_4 with respect to the graphene edges ("direct" or "tilted"). This suggests that different possibilities exist for tuning the electronic characteristics of FeN_4-graphene, which has been recognized as necessary for improving their catalytic activity in ORR [48].

Discussing the property variations induced by the presence of FeN_4 at the edges, we first noted that the spin-unpolarized (non-magnetic) carbon states in the armchair terminated GNRs remained as such also when interacting with the FeN_4 defects at the edges, but the FeN_4-AGNR electronic structure changed from semi-conducting for the FeN_4 integrated in-plane of AGNR to half-metallic for the "direct" FeN_4 integrated at the edge (Figure 4d). Half-metallicity is revealed also by the band structure of the edge-doped ZGNRs (Figure 4b, last column). The spin suppression caused by the introduction of an FeN_4 moiety at the edge of the ZGNR makes its electronic structure comparable with the spin-unpolarized AGNR edges, therefore suppressing the difference in the electronic properties between AGNR and ZGNR terminations. This suggests that the at-edge location of the FeN_4 active site plays a preponderant role over the edge geometries in the electronic structure. Half-metallic electronic structure was found to be at the origin of the highest catalytic activity in FeN_4-graphene for ORR [48]. We can therefore anticipate that edge-doped FeN_4-GNRs would be highly catalytically active in ORR. Adding this to the highest thermodynamic stability of edge-doped ZGNR and AGNR (see Figure 2: points C1, left hand side and point D1, right hand side), edge-doped FeN_4 graphene nanoribbons can be envisioned as the most promising candidates for ORR catalysis. Moreover, since the edge doping removes the effect of the specific geometrical termination on the graphene electronic structures, we can also extrapolate the validity of our findings to the pyrolyzed powder FeN_xC_y catalysts, regardless of the geometrical organizations of their edge carbons.

Furthermore, the spin-down dz^2 electrons in the at-edge FeN_4-ZGNRs are practically localized at the Fermi level (see Figure 5), which would enhance their interaction with the spin-polarized πp^* orbital of O_2 (the ground state of O_2 is a triplet spin state) or the doublet spin-state of OH*. The increase of spin asymmetry (the difference in the population of spin-up and spin-down states) in the band structure of MeN_4-graphene catalysts was also suggested to improve the catalytic activity in the CO oxidation reaction [54]. Following the evolution of the band diagrams in Figure 4b,d, we observed that moving FeN_4 from the in-plane to the edge resulted in increasing the differences between the populations of spin-down and spin-up channels close to the Fermi level, thus highlighting again the edge-doped FeN_4 graphene GNRs as potentially highly active catalysts. Furthermore, the interactions between Cp_z and Fe3d magnetic moments can significantly modify the total magnetic moment of the entire FeN_4-graphene material, as demonstrated in Table 1. The spin polarization of the edge carbons in the various FeN_4-graphene models need to be considered in addition to the Fe spin density in the DFT calculations when accessing the magnetic properties of the active FeN_4 sites.

3. Materials and Methods

Computational Details

All DFT calculations were performed using the Vienna Ab Initio Simulation Package [55,56]. The interactions between electron and nuclei were described within the framework of PAW formalism. The exchange–correlation energy was calculated within the Perdew, Burke, and Ernzerhof formulation of the generalized-gradient approximation (GGA-PBE) [57,58]. After the extensive test calculations for the total energy convergence, the kinetic energy cut-off for plane wave expansion was set to 500 eV. To simulate the periodic graphene nanoribbons (GNRs), the distance between the graphene layers was maintained at 15 Å for edge-edge and layer-layer distances. Structural relaxation is performed with a converge criteria of 1×10^{-2} eV/Å on force and a Γ-centered k-point grid of $15 \times 1 \times 1$ for ZGNR model and $1 \times 15 \times 1$ for AGNR model. Increasing the k-points up to 20 along the periodicity direction did not significantly change the energy values, confirming the adequacy of the chosen grid. An electron self-consistency loop was performed with accurate convergence of the magnetic state by assigning initial spin moments to each atom. We consistently employed the Fermi–Dirac smearing method across all cases, setting the smearing parameter to 0.03. The spin moment distribution of the FeN_4 motif embedded GNRs models was calculated using the Bader analysis as described in ref. [59].

To study the formation of the FeN_4 active sites, the computed formation energies E_f for Fe-N_4-GNR was based on the following equation

$$E_f = \left(\left(E_{Fe-N_4-GNR} + (\mu_C \times N_{subtracted})\right) - (E_{GNR} + (\mu_N \times N_{substituted}) + E_{Fe})\right) \quad (1)$$

where E_{Fe-N_4-GNR} is the total energy of the Fe-N4 substituted model; μ_C is the chemical potential of carbon, defined as the total energy of graphene per carbon atom; E_{GNR} is the total energy of pristine (perfect) GNR; μ_N is the chemical potential of nitrogen, taken as one-half of the total energy of the N_2 molecule in the gas phase, and E_{Fe} is the energy of the isolated Fe atom in the gas phase. In the above equation, the number of subtracted carbon atoms was six (two of them come from double carbon vacancy and the remaining four carbon atoms are subtracted to be replaced by nitrogen) and the number of substituted nitrogen atoms was four. Throughout this article, a lower formation energy signifies a higher probability of FeN_4 formation on a selected position.

Fe binding energy was calculated as

$$E_B = \left(\left(E_{Fe-N_4-GNR}\right) - \left(E_{N4-GNR} + E_{Fe}\right)\right) \quad (2)$$

The binding energies (E_B) were defined as the difference between the energy of the Fe atom and N4 in the substitutional position of the GNRs and the energy of the reconstructed with double vacancy and substituted 4 N atoms (E_{N4-GNR}) plus the energy of isolated Fe atom (E_{Fe}).

The different initial spin moments on the two edges were considered to obtain the energy difference between the AFM and FM configurations for each doped ZGNR; however, unit cells with a single FeN_4 motif allowed only for ferromagnetic solution. Thus, we doubled the unit cell along the x direction ($2 \times 1 \times 1$) to explore the energetically competing antiferromagnetic spin state.

Spin-density distribution analysis was performed using a two-phase computational approach. Initially, charge-density calculations were performed for both core and valence electron shells, and these densities were subsequently integrated to yield the total charge density. Magnetic charge density was derived by subtracting the spin-down charge density from the spin-up charge density. The final phase involved Bader analysis, which partitions the magnetic charge density with reference to the total charge density. The partial charge densities were calculated by taking into account all k-points. Subsequently, using a post-processing bash script, the spin-up and spin-down charge densities were separated from these partial charge densities within the fat-band profile.

4. Conclusions

In this study, we present an evaluation of the thermodynamic stability, electronic, and magnetic properties as a function of the position and orientation of FeN_4, covalently integrated into graphene nanoribbons with zigzag and armchair terminations in view of their potential applications as single-atom electrocatalysts. The DFT calculations with periodic boundary conditions were employed. Our results demonstrate that the covalent integration of FeN_4 defects in both ZGNR and AGNR results in thermodynamically stable materials independent of the doping location. Nevertheless, the differences in the binding and formation energies oscillates within 15%; at-edge doped GNRs are the most energetically stable structures.

The spin-moment distribution in zigzag graphene nanoribbons that integrate FeN_4 is not determined only by the spin polarization of 3d electrons, but also by the spin polarization of the Cp_z electrons. The spin-polarized edge carbons interact with Fe3d spin-polarized states if FeN_4 is located near or at the edge of the ZGNR, which results in suppression of the spin moments of edge carbons. In the latter cases, the Cp_z electrons change their spin orientations, but the Fe3d electrons preserve their spin moments. Consequently, the total magnetic moment of the FeN_4-ZGNRs varies according to the spin orientations on the carbons at the opposite edge and at iron. The spin-unpolarized states of the edge carbons in AGNR remain unaffected by the FeN_4 spin moments.

The spin quench at edge carbons is manifested with the reduction or removal of edge electronic states from Fermi level in the band gap. The FeN_4 at-edge location transforms the electronic structures to half-metallic and increases the spin asymmetry in both ZGNR and AGNR; therefore, integrating FeN_4 at the edges removes the effect of edge geometries on the electronic properties of the catalyst. This suggests that the location of the FeN_4 active sites will be determinant not only for the catalyst stability, but also for its reactivity, being recognized as highly dependent on the FeN_4-graphene magnetic and electronic structures. Since the edge geometries play a minor or non-role in the electronic structure of at-edge FeN_4-graphene, FeN_4-GNRs can be considered as relevant models for describing the magnetic and electronic properties in relation to the reactivity of pyrolyzed Fe-N-C catalysts.

Supplementary Materials: The following supporting information can be downloaded at: https://www.mdpi.com/article/10.3390/molecules29020479/s1, Figure S1: Spin density distributions and sub-band diagrams of FM and AFM ordered ZGNRs; Figure S2: Charge density composition of AFM edges of the undoped ZGNR; Figure S3: Spin-density and sub-band structures in "tilted" FeN_4-AGNR structures; Figure S4: The lattice structure and sub-band diagram of AGNR; Figure S5: Spin density and band diagram of FM and AFM ordered (2 × 1 × 1) supercell of FeN_4-ZGNRs; Figure S6: Spin density and band diagram of FM and AFM ordered (2 × 1 × 1) supercells of FeN4-AGNR; Figure S7: Spin density FM and AFM ordered at-edge FeN_4-ZGNRs; Figure S8: Spin density of at-edge FeN_4 ZGNR (supercell lattice [2 × 1 × 1]).

Author Contributions: Conceptualization, T.M.; I.C.O. performed all the calculations and data curation; T.M. writing—original draft preparation with inputs from the co-authors; I.C.O. and F.J. review and editing. All authors have read and agreed to the published version of the manuscript.

Funding: This research was supported by the LabExCheMISyst ANR-10-LABX-05-01.

Institutional Review Board Statement: Not applicable.

Informed Consent Statement: Not applicable.

Data Availability Statement: Data are contained within the article and supplementary materials.

Acknowledgments: The authors acknowledge the access to the HPC resources of CCRT/CINES/IDRIS, which was granted under the allocation AXXX0807369.

Conflicts of Interest: The authors declare no conflicts of interest.

References

1. Son, Y.W.; Cohen, M.L.; Louie, S.G. Half-Metallic Graphene Nanoribbons. *Nature* **2006**, *444*, 347–349. [CrossRef] [PubMed]
2. Yazyev, O.V. Emergence of Magnetism in Graphene Materials and Nanostructures. *Rep. Prog. Phys.* **2010**, *73*, 056501. [CrossRef]
3. Yazyev, O.V. A Guide to the Design of Electronic Properties of Graphene Nanoribbons. *Acc. Chem. Res.* **2013**, *46*, 2319–2328. [CrossRef] [PubMed]
4. Ejg, S.; Ayuela, A.; Sánchez-Portal, D. First-Principles Study of Substitutional Metal Impurities in Graphene: Structural, Electronic and Magnetic Properties. *New J. Phys.* **2010**, *12*, 053012. [CrossRef]
5. Gao, S.; Yang, L. Edge-Insensitive Magnetism and Half Metallicity in Graphene Nanoribbons. *J. Phys. Condens. Matter* **2018**, *30*, 48LT01. [CrossRef] [PubMed]
6. Ota, N.; Gorjizadeh, N.; Kawazoe, Y. Multiple Spin State Analysis in Radical Carbon Edge and Oxygen Edge Graphene-like Molecules. *J. Magn. Soc. Jpn.* **2011**, *35*, 414–419. [CrossRef]
7. Ota, N.; Gorjizadeh, N.; Kawazoe, Y. Multiple Spin State Analysis of Magnetic Nano Graphene. *J. Magn. Soc. Jpn.* **2011**, *35*, 360–365. [CrossRef]
8. Popov, Z.I.; Mikhaleva, N.S.; Visotin, M.A.; Kuzubov, A.A.; Entani, S.; Naramoto, H.; Sakai, S.; Sorokin, P.B.; Avramov, P.V. The Electronic Structure and Spin States of 2D Graphene/VX2 (X = S, Se) Heterostructures. *Phys. Chem. Chem. Phys.* **2016**, *18*, 33047–33052. [CrossRef] [PubMed]
9. Sarmah, A.; Hobza, P. Understanding the Spin-Dependent Electronic Properties of Symmetrically Far-Edge Doped Zigzag Graphene Nanoribbon from a First Principles Study. *RSC Adv.* **2017**, *7*, 46604–46614. [CrossRef]
10. Wu, F.; Kan, E.; Xiang, H.; Wei, S.H.; Whangbo, M.H.; Yang, J. Magnetic States of Zigzag Graphene Nanoribbons from First Principles. *Appl. Phys. Lett.* **2009**, *94*, 223105. [CrossRef]
11. Zhang, W.X.; He, C.; Li, T.; Gong, S.B. Tuning Electronic and Magnetic Properties of Zigzag Graphene Nanoribbons with a Stone-Wales Line Defect by Position and Axis Tensile Strain. *RSC Adv.* **2015**, *5*, 33407–33413. [CrossRef]
12. Brede, J.; Merino-Díez, N.; Berdonces, A.; Sanz, S.; Domínguez-Celorrio, A.; Lobo-Checa, J.; Vilas-Varela, M.; Peña, D.; Frederiksen, T.; Pascual, J.I.; et al. Detecting the Spin-Polarization of Edge States in Graphene Nanoribbons. *Nat. Commun.* **2023**, *14*, 6677. [CrossRef] [PubMed]
13. de la Torre, B.; Švec, M.; Hapala, P.; Redondo, J.; Krejčí, O.; Lo, R.; Manna, D.; Sarmah, A.; Nachtigallová, D.; Tuček, J.; et al. Non-Covalent Control of Spin-State in Metal-Organic Complex by Positioning on N-Doped Graphene. *Nat. Commun.* **2018**, *9*, 2831. [CrossRef]
14. Kinikar, A.; Xu, X.; Di Giovannantonio, M.; Gröning, O.; Eimre, K.; Pignedoli, C.A.; Müllen, K.; Narita, A.; Ruffieux, P.; Fasel, R. On-Surface Synthesis of Edge-Extended Zigzag Graphene Nanoribbons. *Adv. Mater.* **2023**, *35*, 2306311. [CrossRef]
15. Bundaleska, N.; Dias, A.; Bundaleski, N.; Felizardo, E.; Henriques, J.; Tsyganov, D.; Abrashev, M.; Valcheva, E.; Kissovski, J.; Ferraria, A.M.; et al. Prospects for Microwave Plasma Synthesized N-Graphene in Secondary Electron Emission Mitigation Applications. *Sci. Rep.* **2020**, *10*, 13013. [CrossRef] [PubMed]
16. Yamada, Y.; Tanaka, H.; Kubo, S.; Sato, S. Unveiling Bonding States and Roles of Edges in Nitrogen-Doped Graphene Nanoribbon by X-Ray Photoelectron Spectroscopy. *Carbon* **2021**, *185*, 342–367. [CrossRef]
17. Xia, D.; Yang, X.; Xie, L.; Wei, Y.; Jiang, W.; Dou, M.; Li, X.; Li, J.; Gan, L.; Kang, F. Direct Growth of Carbon Nanotubes Doped with Single Atomic Fe–N4 Active Sites and Neighboring Graphitic Nitrogen for Efficient and Stable Oxygen Reduction Electrocatalysis. *Adv. Funct. Mater.* **2019**, *29*, 1906174. [CrossRef]
18. Boukhvalov, D.W.; Katsnelson, M.I. Chemical Functionalization of Graphene with Defects. *Nano Lett.* **2008**, *8*, 4374–4379. [CrossRef]
19. Liao, M.S.; Scheiner, S. Electronic Structure and Bonding in Metal Porphyrins, Metal = Fe, Co, Ni, Cu, Zn. *J. Chem. Phys.* **2002**, *117*, 205–219. [CrossRef]
20. Yu, G.; Lü, X.; Jiang, L.; Gao, W.; Zheng, Y. Structural, Electronic and Magnetic Properties of Transition-Metal Embedded Zigzag-Edged Graphene Nanoribbons. *J. Phys. D Appl. Phys.* **2013**, *46*, 375303. [CrossRef]
21. Kattel, S. Magnetic Properties of 3d Transition Metals and Nitrogen Functionalized Armchair Graphene Nanoribbon. *RSC Adv.* **2013**, *3*, 21110–21117. [CrossRef]
22. Jiang, R.; Qiao, Z.; Xu, H.; Cao, D. Novel 2D Carbon Material T-Graphene Supported 3d Transition Metal Single Atoms as Efficient Oxygen Reduction Catalysts. *Nanoscale* **2023**, *15*, 16775–16783. [CrossRef] [PubMed]
23. Shang, Y.; Zhao, J.X.; Wu, H.; Cai, Q.H.; Wang, X.G.; Wang, X.Z. Chemical Functionalization of Pyridine-like and Porphyrin-like Nitrogen-Doped Carbon (CNx) Nanotubes with Transition Metal (TM) Atoms: A Theoretical Study. *Theor. Chem. Acc.* **2010**, *127*, 727–733. [CrossRef]
24. Chan, K.T.; Neaton, J.B.; Cohen, M.L. First-Principles Study of Metal Adatom Adsorption on Graphene. *Phys. Rev. B Condens. Matter Mater. Phys.* **2008**, *77*, 235430. [CrossRef]
25. Longo, R.C.; Carrete, J.; Ferrer, J.; Gallego, L.J. Structural, Magnetic, and Electronic Properties of Nin and Fen Nanostructures (N = 1–4) Adsorbed on Zigzag Graphene Nanoribbons. *Phys. Rev. B Condens. Matter Mater. Phys.* **2010**, *81*, 115418. [CrossRef]
26. Zheng, S.; Yan, X.; Yang, Y.; Xu, J. Identifying Structure-Property Relationships through SMILES Syntax Analysis with Self-Attention Mechanism. *J. Chem. Inf. Model.* **2019**, *59*, 914–923. [CrossRef]
27. Li, Y.; Zhou, Z.; Shen, P.; Chen, Z. Spin Gapless Semiconductor-Metal-Half-Metal Properties in Nitrogen-Doped Zigzag Graphene Nanoribbons. *ACS Nano* **2009**, *3*, 1952–1958. [CrossRef]

28. Holby, E.F.; Taylor, C.D. Control of Graphene Nanoribbon Vacancies by Fe and N Dopants: Implications for Catalysis. *Appl. Phys. Lett.* **2012**, *101*, 064102. [CrossRef]
29. Luo, F.; Roy, A.; Silvioli, L.; Cullen, D.A.; Zitolo, A.; Sougrati, M.T.; Oguz, I.C.; Mineva, T.; Teschner, D.; Wagner, S.; et al. Author Correction: P-Block Single-Metal-Site Tin/Nitrogen-Doped Carbon Fuel Cell Cathode Catalyst for Oxygen Reduction Reaction (Nature Materials, (2020), 19, 11, (1215-1223), 10.1038/S41563-020-0717-5). *Nat. Mater.* **2023**, *22*, 146. [CrossRef]
30. Kaiser, S.K.; Chen, Z.; Faust Akl, D.; Mitchell, S.; Pérez-Ramírez, J. Single-Atom Catalysts across the Periodic Table. *Chem. Rev.* **2020**, *120*, 11703–11809. [CrossRef]
31. Kattel, S.; Atanassov, P.; Kiefer, B. A Density Functional Theory Study of Oxygen Reduction Reaction on Non-PGM Fe-Nx-C Electrocatalysts. *Phys. Chem. Chem. Phys.* **2014**, *16*, 13800–13806. [CrossRef]
32. Zitolo, A.; Goellner, V.; Armel, V.; Sougrati, M.T.; Mineva, T.; Stievano, L.; Fonda, E.; Jaouen, F. Identification of Catalytic Sites for Oxygen Reduction in Iron- and Nitrogen-Doped Graphene Materials. *Nat. Mater.* **2015**, *14*, 937–942. [CrossRef] [PubMed]
33. Zitolo, A.; Ranjbar-Sahraie, N.; Mineva, T.; Li, J.; Jia, Q.; Stamatin, S.; Harrington, G.F.; Lyth, S.M.; Krtil, P.; Mukerjee, S.; et al. Identification of Catalytic Sites in Cobalt-Nitrogen-Carbon Materials for the Oxygen Reduction Reaction. *Nat. Commun.* **2017**, *8*, 957. [CrossRef]
34. Ju, W.; Bagger, A.; Hao, G.P.; Varela, A.S.; Sinev, I.; Bon, V.; Roldan Cuenya, B.; Kaskel, S.; Rossmeisl, J.; Strasser, P. Understanding Activity and Selectivity of Metal-Nitrogen-Doped Carbon Catalysts for Electrochemical Reduction of CO_2. *Nat. Commun.* **2017**, *8*, 944. [CrossRef] [PubMed]
35. Xie, Y.; Yang, Y.; Yang, W.; Liu, N.; Chen, X. The Relationship between the Local Environment, N-Type, Spin State and Catalytic Functionality of Carbon-Hosted FeII/III-N4 for the Conversion of CO_2 to CO. *Phys. Chem. Chem. Phys.* **2023**, *25*, 18889–18902. [CrossRef] [PubMed]
36. Li, X.F.; Li, Q.K.; Cheng, J.; Liu, L.; Yan, Q.; Wu, Y.; Zhang, X.H.; Wang, Z.Y.; Qiu, Q.; Luo, Y. Conversion of Dinitrogen to Ammonia by FeN3-Embedded Graphene. *J. Am. Chem. Soc.* **2016**, *138*, 8706–8709. [CrossRef] [PubMed]
37. Kattel, S.; Wang, G. A Density Functional Theory Study of Oxygen Reduction Reaction on Me-N 4 (Me = Fe, Co, or Ni) Clusters between Graphitic Pores. *J. Mater. Chem. A Mater.* **2013**, *1*, 10790–10797. [CrossRef]
38. Fei, H.; Dong, J.; Feng, Y.; Allen, C.S.; Wan, C.; Volosskiy, B.; Li, M.; Zhao, Z.; Wang, Y.; Sun, H.; et al. General Synthesis and Definitive Structural Identification of MN4C4 Single-Atom Catalysts with Tunable Electrocatalytic Activities. *Nat. Catal.* **2018**, *1*, 63–72. [CrossRef]
39. Ma, J.; Wang, L.; Deng, Y.; Zhang, W.; Wu, T.; Song, Y. Mass production of high-performance single atomic FeNC electrocatalysts via sequenced ultrasonic atomization and pyrolysis process. *Sci. China Mater.* **2021**, *64*, 631–641. [CrossRef]
40. Saputro, A.G.; Kasai, H. Oxygen Reduction Reaction on Neighboring Fe-N4 and Quaternary-N Sites of Pyrolized Fe/N/C Catalyst. *Phys. Chem. Chem. Phys.* **2015**, *17*, 3059–3071. [CrossRef]
41. Sougrati, M.T.; Goellner, V.; Schuppert, A.K.; Stievano, L.; Jaouen, F. Probing Active Sites in Iron-Based Catalysts for Oxygen Electro-Reduction: A Temperature-Dependent 57Fe Mössbauer Spectroscopy Study. *Catal. Today* **2016**, *262*, 110–120. [CrossRef]
42. Mineva, T.; Matanovic, I.; Atanassov, P.; Sougrati, M.T.; Stievano, L.; Clémancey, M.; Kochem, A.; Latour, J.M.; Jaouen, F. Understanding Active Sites in Pyrolyzed Fe-N-C Catalysts for Fuel Cell Cathodes by Bridging Density Functional Theory Calculations and 57Fe Mössbauer Spectroscopy. *ACS Catal.* **2019**, *9*, 9359–9371. [CrossRef]
43. Li, J.; Sougrati, M.T.; Zitolo, A.; Ablett, J.M.; Oğuz, I.C.; Mineva, T.; Matanovic, I.; Atanassov, P.; Huang, Y.; Zenyuk, I.; et al. Identification of Durable and Non-Durable FeNx Sites in Fe–N–C Materials for Proton Exchange Membrane Fuel Cells. *Nat. Catal.* **2021**, *4*, 10–19. [CrossRef]
44. Szakacs, C.E.; Lefèvre, M.; Kramm, U.I.; Dodelet, J.P.; Vidal, F. A Density Functional Theory Study of Catalytic Sites for Oxygen Reduction in Fe/N/C Catalysts Used in H_2/O_2 Fuel Cells. *Phys. Chem. Chem. Phys.* **2014**, *16*, 13654–13661. [CrossRef]
45. Kramm, U.I.; Abs-Wurmbach, I.; Herrmann-Geppert, I.; Radnik, J.; Fiechter, S.; Bogdanoff, P. Influence of the Electron-Density of FeN[Sub 4]-Centers Towards the Catalytic Activity of Pyrolyzed FeTMPPCl-Based ORR-Electrocatalysts. *J. Electrochem. Soc.* **2011**, *158*, B69. [CrossRef]
46. Wagner, S.; Auerbach, H.; Tait, C.E.; Martinaiou, I.; Kumar, S.C.N.; Kübel, C.; Sergeev, I.; Wille, H.C.; Behrends, J.; Wolny, J.A.; et al. Elucidating the Structural Composition of an Fe–N–C Catalyst by Nuclear- and Electron-Resonance Techniques. *Angew. Chem. Int. Ed.* **2019**, *58*, 10486–10492. [CrossRef] [PubMed]
47. Li, X.F.; Liu, L.; Yan, Q.; Li, Q.K.; Wang, Y.; Deng, M.; Qiu, Q. Strong Current Polarization and Perfect Negative Differential Resistance in Few-FeN4-Embedded Zigzag Graphene Nanoribbons. *Phys. Chem. Chem. Phys.* **2017**, *19*, 2674–2678. [CrossRef]
48. Wu, L.; Cao, X.; Hu, W.; Ji, Y.; Zhu, Z.Z.; Li, X.F. Improving the Oxygen Reduction Reaction Activity of FeN4-Graphene via Tuning Electronic Characteristics. *ACS Appl. Energy Mater.* **2019**, *2*, 6634–6641. [CrossRef]
49. Cervantes-Sodi, F.; Csányi, G.; Piscanec, S.; Ferrari, A.C. Edge-Functionalized and Substitutionally Doped Graphene Nanoribbons: Electronic and Spin Properties. *Phys. Rev. B Condens. Matter Mater. Phys.* **2008**, *77*, 165427. [CrossRef]
50. Holby, E.F.; Wu, G.; Zelenay, P.; Taylor, C.D. Structure of Fe − N. *J. Phys. Chem. C* **2014**, *118*, 14388–14393.
51. Yu, S.S.; Zheng, W.T.; Wen, Q.B.; Jiang, Q. First Principle Calculations of the Electronic Properties of Nitrogen-Doped Carbon Nanoribbons with Zigzag Edges. *Carbon* **2008**, *46*, 537–543. [CrossRef]
52. Hu, X.; Zhang, W.; Sun, L.; Krasheninnikov, A. V Gold-Embedded Zigzag Graphene Nanoribbons as Spin Gapless Semiconductors. *Phys. Rev. B* **2012**, *86*, 195418. [CrossRef]

53. Hammer, B.; Nørskov, J.K. Theoretical Surface Science and Catalysis—Calculations and Concepts. In *Advances in Catalysis*; Academic Press: Cambridge, MA, USA, 2000; Volume 45, pp. 71–129. ISBN 0360-0564.
54. Groome, C.; Ngo, H.; Li, J.; Wang, C.S.; Wu, R.; Ragan, R. Influence of Magnetic Moment on Single Atom Catalytic Activation Energy Barriers. *Catal. Lett.* **2022**, *152*, 1347–1357. [CrossRef]
55. Kresse, G.; Furthmüller, J. Efficient Iterative Schemes for Ab Initio Total-Energy Calculations Using a Plane-Wave Basis Set. *Phys. Rev. B* **1996**, *54*, 11169–11186. [CrossRef]
56. Kresse, G.; Furthmüller, J. Efficiency of Ab-Initio Total Energy Calculations for Metals and Semiconductors Using a Plane-Wave Basis Set. *Comput. Mater. Sci.* **1996**, *6*, 15–50. [CrossRef]
57. Perdew, J.P.; Burke, K.; Ernzerhof, M. Erratum: Generalized Gradient Approximation Made Simple (Physical Review Letters (1996) 77 (3865)). *Phys. Rev. Lett.* **1997**, *78*, 1396. [CrossRef]
58. Perdew, J.P.; Burke, K.; Ernzerhof, M. Generalized Gradient Approximation Made Simple. *Phys. Rev. Lett.* **1996**, *77*, 3865–3868. [CrossRef]
59. Tang, S.; Case, D.A. Vibrational Averaging of Chemical Shift Anisotropies in Model Peptides. *J. Biomol. NMR* **2007**, *38*, 255–266. [CrossRef]

molecules

MDPI

Article

Modulating Optoelectronic and Elastic Properties of Anatase TiO_2 for Photoelectrochemical Water Splitting

Akbar Hussain [1], Abdur Rauf [1], Ejaz Ahmed [2], Muhammad Saleem Khan [3], Shabeer Ahmad Mian [1,*] and Joonkyung Jang [4,*]

1 Department of Physics, University of Peshawar, Peshawar 25120, Pakistan
2 Department of Physics, Abdul Wali Khan University, Mardan 23200, Pakistan
3 Department of Chemical Engineering, NFC Institute of Engineering & Technology, Multan 60000, Pakistan
4 Department of Nano Energy Engineering, Pusan National University, Busan 46241, Republic of Korea
* Correspondence: shabeerahmad@uop.edu.pk (S.A.M.); jkjang@pusan.ac.kr (J.J.)

Abstract: Titanium dioxide (TiO_2) has been investigated for solar-energy-driven photoelectrical water splitting due to its suitable band gap, abundance, cost savings, environmental friendliness, and chemical stability. However, its poor conductivity, weak light absorption, and large indirect bandgap (3.2 eV) has limited its application in water splitting. In this study, we precisely targeted these limitations using first-principle techniques. TiO_2 only absorbs near-ultraviolet radiation; therefore, the substitution (2.1%) of Ag, Fe, and Co in TiO_2 significantly altered its physical properties and shifted the bandgap from the ultraviolet to the visible region. Cobalt (Co) substitution in TiO_2 resulted in high absorption and photoconductivity and a low bandgap energy suitable for the reduction in water without the need for external energy. The calculated elastic properties of Co-doped TiO_2 indicate the ductile nature of the material with a strong average bond strength. Co-doped TiO_2 exhibited fewer microcracks with a mechanically stable composition.

Keywords: optoelectronic; elastic moduli; transition metals; DFT; NHE; water splitting

1. Introduction

Overcoming the energy crisis without affecting the environment is a crucial challenge for scientists and engineers in the near future. For this purpose, various techniques have been proposed, and the utilization of absorbed sunlight energy through photo/electrocatalysis is an especially promising solution. Honda and Fujishima achieved water splitting using a TiO_2-based photo/electrocatalyst. TiO_2 exists in three phases: anatase, brookite, and rutile [1]. They are all used in photo-electrochemical water-splitting applications [2]. Among them, TiO_2 anatase has been hailed as one of the most promising candidates due to its indirect bandgap [3]. In contrast, brookite and rutile exhibited a direct bandgap of 3.0 and 3.5 eV [2,4,5]. Photo-excited electrons have a longer lifetime in indirect bandgap semiconductors than in direct bandgap semiconductors [2,6] because photo-excited electrons cannot migrate directly from the conduction band (CB) to the valence band (VB) [2], decreasing the charge carrier recombination rate. Additionally, due to the rapid migration of charge carriers [2], TiO_2 anatase had a lower charge-carrier-recombination rate and the smallest average effective mass [7] of photo-generated electrons, as compared to the brookite and rutile phases; therefore, the superior photo-electrochemical water-splitting properties exhibited by TiO_2 anatase were not surprising.

Semiconductors such as hematite Fe_2O_3 [8–10], tungsten trioxide WO_3 [11–13], and TiO_2 [14–16], as well as perovskites including ABX_3 [17], $SrTiO_3$ [18], metal oxides [19], and metal sulfides [20], have been used for this purpose. However, due to its low cost of production [21], easy availability, non-toxicity [22], and thermodynamic stability, TiO_2 anatase has been investigated meticulously [23,24]. Modern industrialization, technological development, and the rapidly growing global population have caused higher energy consumption, which has resulted in severe environmental pollution due to the burning of hydrocarbons. Clean energy is required to achieve a green environment. Hydrogen is a renewable source, and water is available in an unlimited supply. Photo-electrochemical water splitting resulting in hydrogen fuel and oxygen gas by utilizing sunlight through the application of photo-electrode materials has considerable technological potential [25]. The nature of the delocalized photo-excited holes and electrons, which are generated because of the absorption of light photons of appropriate energy, could play a vital role in the application of water for redox reactions to produce hydrogen and oxygen. The reaction of water with the photo-anode surface occurs in the following manner.

$$H_2O + 2e^+ \rightarrow 2H^+ + \frac{1}{2}O_2 \quad (1)$$

After the oxidation of water, the electrons are transferred to the cathode using an external circuit, while H^+ moves toward the cathode within the electrolyte, where H^+ and e^- react in the following manner:

$$2H^+ + 2e^- \rightarrow H_2 \quad (2)$$

Pure TiO_2 anatase absorbs only ultraviolet light (3–4% of the solar spectrum) due to its wide (3.2 eV) bandgap [3]. Furthermore, it is extremely easy for photo-generated holes and electrons to recombine into pure TiO_2 anatase, which reduces the photo-electrochemical water-splitting phenomena [26]. Therefore, only two problems have been encountered during this enhanced photo-electrochemical water splitting: (1) a shrinking of the bandgap and (2) a reduction in the charge-carrier-recombination rate. Various existing techniques, such as doping, surface modification, and the construction of heterojunctions, have been proposed to overcome these flaws [27,28]. Among them, doping techniques have significantly improved the optical, electrical, and photocatalytic performance of TiO_2. Using the developed methodology, the aforementioned issues have been addressed with nonmetallic (N, S, P, B, and C [29,30]) and metallic (Fe, Mo, Au, Rh, and Cu) dopants [31]. Giovanni et al. investigated the N-doping effect on anatase surfaces using first-principle calculations and realized enhanced light absorption because of the creation of oxygen vacancy [23]. Xiaoye et al. studied metal-doped anatase and reported that Cu-doping significantly enhanced its photocatalytic activity in PEC applications. Furthermore, Anindita et al. performed thermoelectric mismatching and bandgap calculations for anatase thin films [24]. Illyas et al. comprehensively studied the optical and electrical properties of TiO_2 rutile and found that Co-doped TiO_2 rutile absorbed a large spectrum of solar light. To obtain sufficient knowledge regarding the interatomic and other solid-state phenomena, understanding the elastic properties are crucial. Due to the super-hard nature of TiO_2 anatase, it has become a prevalent interest of researchers; therefore, the calculations of the mechanical properties are as vital as those of the electronic and optical properties [32].

Herein, the anatase phase of TiO_2 was comprehensively studied using density-functional-theory (DFT) calculations for PEC water-splitting applications. We thoroughly investigated the impact of Ag, Fe, and Co dopants on the electronic and optical properties of pristine anatase using a DFT + U approach. In addition, the elastic constants (Cij), as well as the shear (G) and bulk (B) moduli, were probed for multifarious fundamental solid-state physical characteristics, including Young's modulus (Y), the Poisson coefficient (υ), and anisotropy (A^U).

Computational Methods and Details

We performed our calculations using DFT with the base variable as the density of electrons, instead of the wave function implemented in SIESTA [33–35]. The functionals of the exchange correlation GGA with exchange correlation of revised the Pardew–Burke–Ernzerhof (RPBE) [36,37] for geometry relation were used. A $2 \times 2 \times 1$ supercell was modeled with 48 atoms to perform the calculations. Furthermore, $6 \times 6 \times 6$ k-points were used for the optical properties, and 3 × 3 × 1 k-points were used for the geometry optimization. A cutoff energy of 200 Ry was selected for all structural optimizations as well as for electronic and optical property calculations. Due to the presence of the strongly correlated d-orbital of Ti, we used an effective U-value of 3.5 eV of the Hubbard model (RPBE+U). Herein, a pseudo-atomic orbital (PAO) basis set with a double-zeta potential (DZP) was assigned to all the atoms in the configuration. In addition, we performed all the calculations using the GGA+U method. Elastic property calculations were carried out using the CASTEP code by considering the GGA proposed by PBE with a 340 eV cut-off plane wave basis set. Additionally, all optical properties' calculations, including absorption coefficients, reflectivity, and energy loss functions were obtained from [38–40] in the interval from 250 to 700 nm.

2. Results and Discussion

2.1. Geometrical Analysis

The effects of doping (Ag, Co, and Fe) on the tetragonal structure of TiO_2 anatase were explored using DFT. The defects produced by the dopants in the host supercell matrix were simulated. The defect geometries were obtained by substituting the Ti atoms with each dopant individually. The optimized geometries of the doped and pure TiO_2 anatases are given in Figure 1.

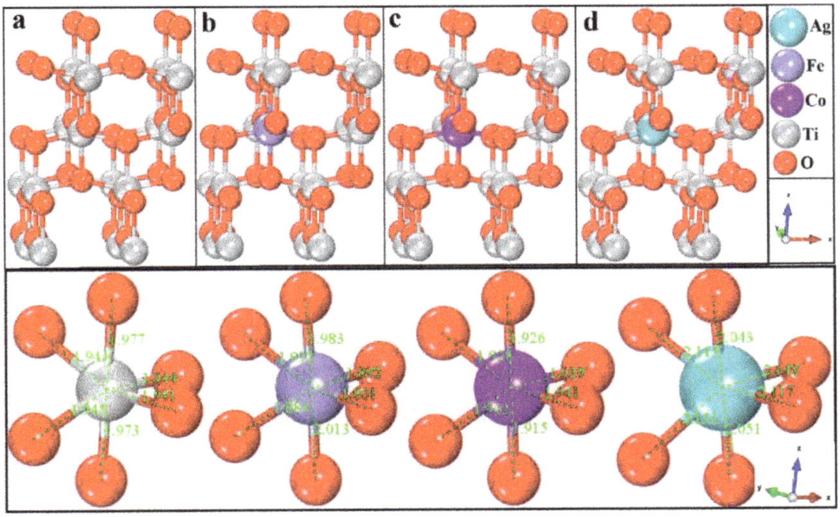

Figure 1. Relaxed geometries of (**a**) TiO_2, (**b**) TiO_2/Fe, (**c**) TiO_2/Co, and (**d**) TiO_2/Ag.

The lattice vectors estimated at the lowest energies are listed in Table 1. The bond lengths of Ti and each dopant with O atoms were calculated and found to be in good agreement with the experimental data and, therefore, deemed favorable for further study. The incorporation of Fe and Co resulted in decreased bond lengths, which caused the cell volume to decrease, as shown in Table 1. Furthermore, the Ag dopant expanded the cell volume due to the long Ag–O bond length. The thermodynamic stability of each doped TiO_2 sample was calculated using the formation energy of pristine and doped TiO_2, using the following equations:

$$E_F = \frac{1}{N}\left(E_T - \sum_{i=\text{Ti,O,Ag,Co,Fe}} n_i E_i\right) \tag{3}$$

where E_T represents the total energy of the preferred system consisting of N atoms. The variable E_i demonstrates the total energy of an isolated i (Ti, O, Ag, Co, Fe) atom, and n_i is the total number of a specific atom i per unit cell. Our estimated formation energy for pristine and Fe-, Co-, and Ag-doped TiO_2 were −9.534, −9.444, −9.418, and −9.245 eV, respectively. Higher negative values of the formation energy indicated their favorable and easy synthesis in the laboratory; this was consistent with the data in the literature. The comparative formation energies of doped and pristine TiO_2 are shown in Figure 2.

Table 1. Lattice parameters, volume, and c/a of pristine and doped TiO_2 anatase.

Type	a (Å)	c(Å)	c/a	V (Å)^3
TiO_2-Pure	3.791	9.510	2.51	
DFT [41]	3.737, 3.741	9.981, 9.964	2.57, 2.66	549.537
Experiment [42]	3.785	9.514	2.51	
TiO_2/Ag	3.821	9.588	2.51	559.621
DFT [43]	3.905	9.825	2.52	
TiO_2/Co	3.802	9.447	2.48	545.971
TiO_2/Fe	3.785	9.496	2.51	545.984
DFT [44]	3.771	9.489	2.49	

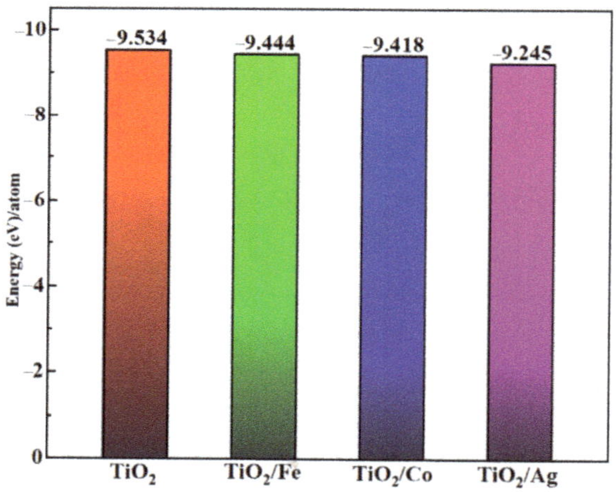

Figure 2. Formation energies per atom of pure and doped TiO_2.

2.2. Electronic Properties

To understand the electronic and optical properties of the pristine and doped compositions, an investigation of the electronic band structure was of prime importance. The band structures of pristine and doped TiO_2 in the SIESTA code were calculated using the GGA+U approximation. All band-structure calculations were performed using high-symmetry points along special lines in the k-space. Figure 3 depicts the band structures of the pristine and doped systems. As shown in Figure 3, TiO_2 was an indirect bandgap (3.18 eV) semiconductor, which was in good agreement with the experimental results [41].

Furthermore, TiO_2/Ag, TiO_2/Co, and TiO_2/Fe showed direct bandgaps of 1.50, 2.02, and 1.58 eV, respectively, along the high-symmetry directions of the Brillouin zones and were in good agreement with another work, as shown in Table 2 [43,44].

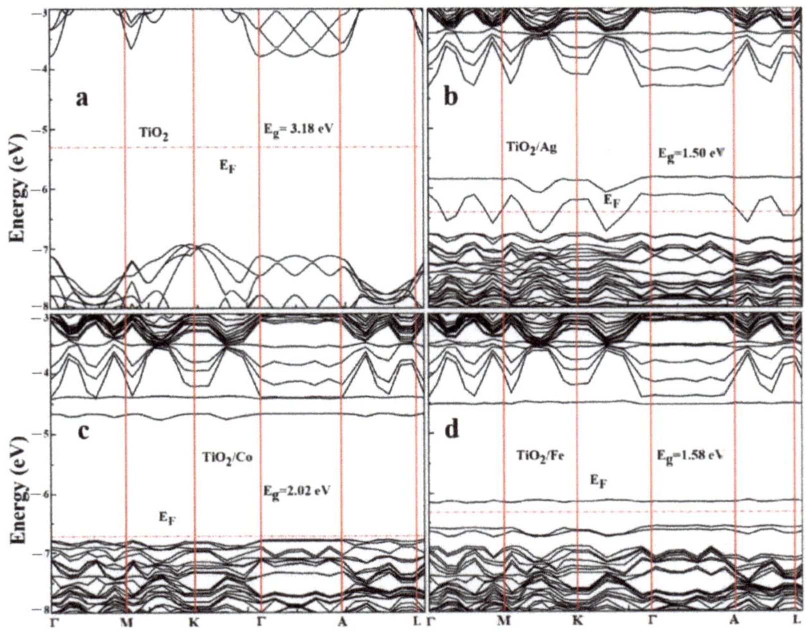

Figure 3. Band structure of (**a**) pristine and (**b**) Ag-, (**c**) Co-, and (**d**) Fe-doped TiO_2.

Table 2. Comparative bandgaps of pristine and doped TiO_2 anatase.

Materials	This Work	Other DFT	Experimental
TiO_2	3.18 eV	3.3 eV [45]	3.2 eV [42]
TiO_2/Ag	1.50 eV	0.9 eV [43]	—
TiO_2/Co	2.02 eV	—	—
TiO_2/Fe	1.58 eV	1.74 eV [44]	—

The incorporation of the dopants (Ag, Co, and Fe) in TiO_2 caused the bandgaps for each compound to shrink, which resulted in the maximum absorption of the solar spectrum, tumbling the rate of recombination and increasing the generation of electron–hole pairs for efficient PEC activity. This reduction occurred in the bandgaps because of the generation of novel energy states near the Fermi level.

To unravel the composition of the valence and conduction bands, the projected density of the states (PDOS) for both the pristine and doped TiO_2 were calculated, as portrayed in Figure 4. For comparison, the PDOS of pristine TiO_2 is also shown in Figure 4a, in which the valance band edge was mostly composed of O-2p states, whereas the conduction band edge was attributed to the unoccupied Ti-3d atomic orbitals. In the case of the Ag dopant, the new states generated by the Ag-3d atomic orbital occurred in the valence-band edge with the overlapping of O-2p states. Similarly, when a Ti atom was substituted by a Co atom, it induces new states of the Co-3d orbitals mixed in with the O-2p and Ti-3d states in the valence band. For the Fe substitution of Ti, the Fe-3d states occurred in the valence band, causing a redshift of absorption and enhancing the photo-response in the visible region. Conclusively, the substitutional impurities in titania caused the shrinkage of the bandgap due to the creation of novel states in the mid-gap, thus reducing the energy barrier in the photon-absorption process, which was the primary condition for high-efficiency photo-anodes in the PEC process.

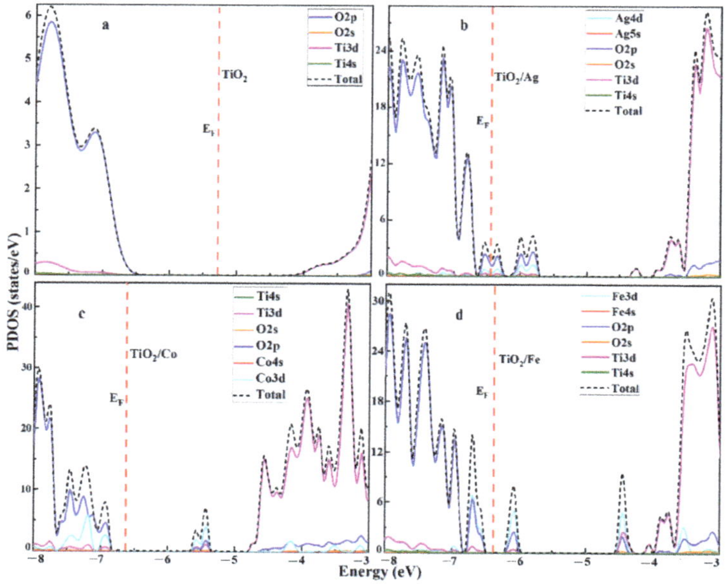

Figure 4. Projected density of states of (**a**) pristine and (**b**) Ag-, (**c**) Co-, and (**d**) Fe-doped TiO_2.

The generation of H_2 via water splitting has been of great importance [46]. To induce water splitting without external energy, the positions of the valence and conduction bands had to be more positive and negative than the oxidation and reduction potentials of the water, respectively. The oxidation potential of water is +1.23 V while its reduction potential is +0 V vs. NHE [47]. In photocatalytic processes, charge separation is a vital factor. If successful charge separation occurs, these charges move to the semiconductor surface and participate in the oxidation and reduction processes. Figure 5 illustrates the energies of the VBM and the CBM of pristine and doped anatase for efficient photocatalytic properties [15]. However, Figure 5 shows that the CBM of anatase was more negative than the water oxidation potential, whereas the VBM edge was more positive, indicating its superiority for PEC activity.

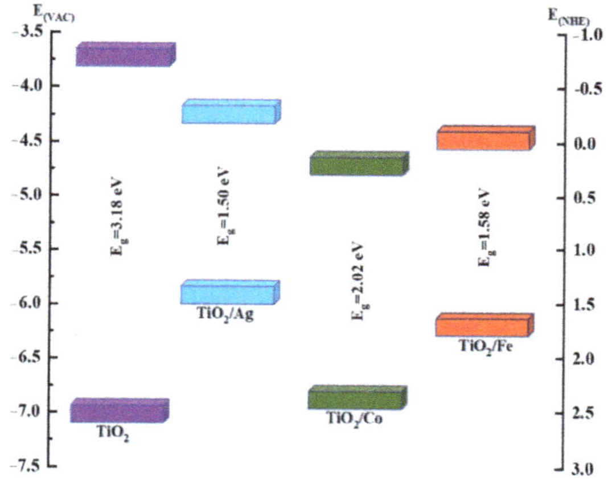

Figure 5. Band-edge comparisons of pure and doped TiO_2 anatase in electron volts. The energy scale is represented in (eV) using either the normal hydrogen-electrode potential (NHE) or the vacuum level as a reference.

The Ag-doped anatase had a suitable conduction-band position for the water reduction reaction, whereas Co-doped anatase had a suitable valence-band position for the water oxidation reaction. Specifically, all dopants significantly improved the VBM position, representing better and promising oxidation reactions for the water-splitting half-reaction. The shift in the CBM and the VBM occurred because of the transition-metal dopants, which contributed to the new energy states.

2.2.1. Real and Imaginary Parts of Dielectric Functions

Dielectric functions define the electronic properties of materials under the effect of incident radiation; they are given by $\varepsilon(\omega) = \varepsilon_1(\omega) + i\ \varepsilon_2(\omega)$. The real part "$\varepsilon_1(\omega)$" demonstrates the material polarizability, whereas the imaginary part $\varepsilon_2(\omega)$ describes the electronic absorption of the material when illuminated by a certain incident photon energy. The calculated values of the real and imaginary parts of the pristine and doped TiO_2 are shown in Figure 6. For pristine TiO_2, ε_1 had a peak value of 5.11 at 387 nm, and the incorporation of the dopants increased this value up to 6.14 at the same wavelength, suggesting the maximum electronic polarization for energy storage devices. Among all the dopants, Co exhibited a high polarizability near 387 nm, as shown in Figure 6a. In Figure 6b, the imaginary part ε_2 exhibited a first peak at 317 nm for pristine TiO_2. The first edge of each material was associated with the fundamental bandgap E_g, which represents the transition between the valence-band maximum (VBM) and the conduction-band minimum (CBM). In the case of pristine TiO_2, the transition of electrons occurred from the O-2p-occupied state (VBM) to the Ti3d unoccupied state (CBM) near the Fermi level in the ultraviolet region, suggesting a higher energy loss than that of the doped TiO_2. Furthermore, for the Co-doped anatase, the peak value was observed at 300 nm. In general, the substitutional doping improved the real and imaginary parts of the dielectric functions in the overall solar spectrum, with considerable advancement in the visible region.

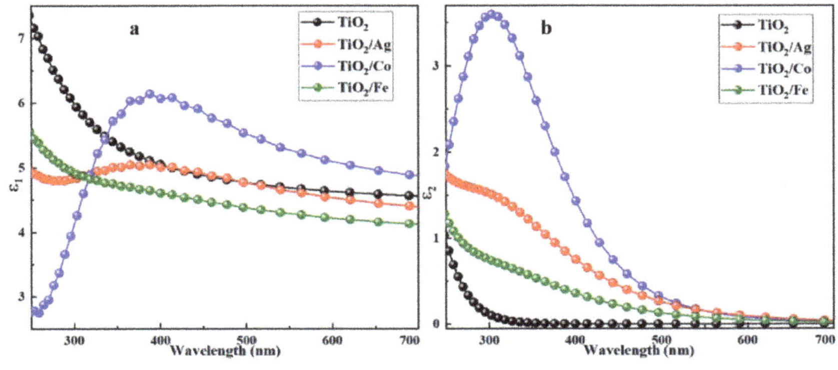

Figure 6. (**a**) Real and (**b**) imaginary parts of the dielectric function for pure and transition-metal-doped TiO_2.

2.2.2. Optical Absorption and Conductivity

To describe the effect of dopants (Ag, Co, and Fe) on the optical absorption and conductivity, we first calculated the two parameters for pristine and doped TiO_2, as portrayed in Figure 7. Pristine TiO_2 had a wide bandgap (3.18 eV) and could only absorb ultraviolet light [42]. The substitutional doping significantly improved the absorption peak in the visible region, showing much better and more promising material properties for photoelectrochemical water-splitting applications. The improvements in the absorption edge, as shown in Figure 7a, were attributed to the narrow bandgap of the pristine TiO_2, and the existence of the new mediator energy states between the VBM and the CBM had decreased the electronic excitation energy needed. Such states were capable of absorbing low-energy photons and generating electron–hole pairs, both of which increased the photoconversion efficiency of the materials. Our computed optical absorption properties for iron (Fe)-doped TiO_2 were consistent with theoretical and experimental data [48]. Among all the dopants, Co showed a substantially larger absorption than the other dopants due to the high density of the occupied and unoccupied states, which utilized low-energy photons for the transition from the VBM to the CBM.

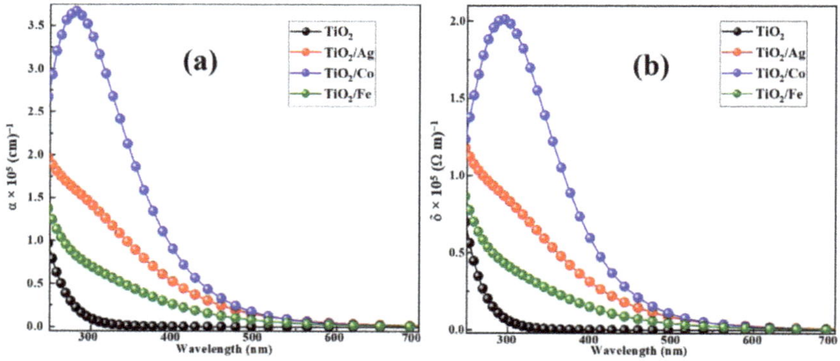

Figure 7. (**a**) Absorption coefficient and (**b**) optical conductivity of pure and metal doped TiO_2 anatase.

In addition, the enhancements of the optical conductivity played a vital role in water splitting and opto-electronic devices, which depend solely on the absorption and refractive index. If a material absorbs greater photon energy than its bandgap, electron–hole pair generation occurs. These pairs move freely in the crystal, which induces optical conductivity. Optical conductivity does not contribute to electrical conductivity because of electronic charge neutrality [49]. For semiconductors and insulators, the electrical conductivity has always been negligible; however, the optical conductivity had been finite due to the optical bandgap. Figure 7b illustrates that the incorporation of the dopants in TiO_2 increased its conductivity, with the largest (by a considerable margin) peak in Co-doped anatase. This enhancement would increase the water reduction reactions due to a redshift in the bandgap energy, which would result in the maximum absorption of the solar spectrum. Optical conductivity followed the same trend as the absorption coefficient.

The calculated reflectance and refractive index for both pristine and transition-metal-doped TiO_2 as a function of the wavelength is shown in Figure 8. The refractive index determines the speed of light propagation in a medium. For a given material, the maximum value of the refractive index indicates that the light slowly propagates within the material; the opposite is true for the minimum refractive index. For doped TiO_2, the refractive index showed a decreasing trend at higher wavelengths and an increasing trend at lower wavelengths. The decreasing trend was due to the optical dispersion of the materials [50]. Cobalt-doped TiO_2 exhibited a maximum refractive index value of 2.51 at $n = 380$ nm, as portrayed in Figure 8.

Figure 8a demonstrates the estimated values of the reflectance for the materials. The reflectivity for Co-doped TiO_2 exhibited a maximum value of approximately 20%. In addition, the pristine and Fe- and Ag-doped TiO_2 showed < 15% reflectivity in the visible range, suggesting their suitability in antireflective-coating applications [51]. All the compositions with weak absorption and conduction exhibited stronger reflectance.

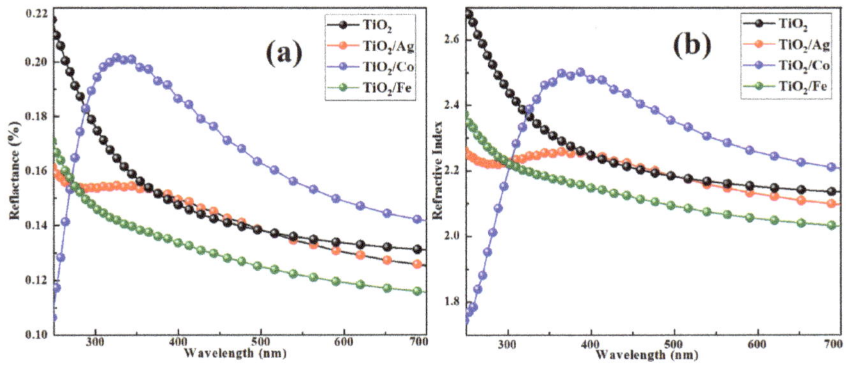

Figure 8. (**a**) Reflectance and (**b**) refractive index for pure and doped TiO_2 anatase.

The loss spectrum was another optical parameter that revealed the spectrum loss during its propagation through the medium. When electrons moved inside the solids, inelastic scattering occurred, which was related to the energy-loss function. Figure 9 displays the energy-loss spectrum as a function of the wavelength from the VBM to the CBM. In addition, the loss spectrum indicated the trailing edges of the reflection spectrum. The highest energy-loss peak characterized the plasma frequency and related resonance. Furthermore, Co-doped TiO_2 showed a sharp decrease, as compared to the other dopants, indicating a rapid decrease in reflectivity.

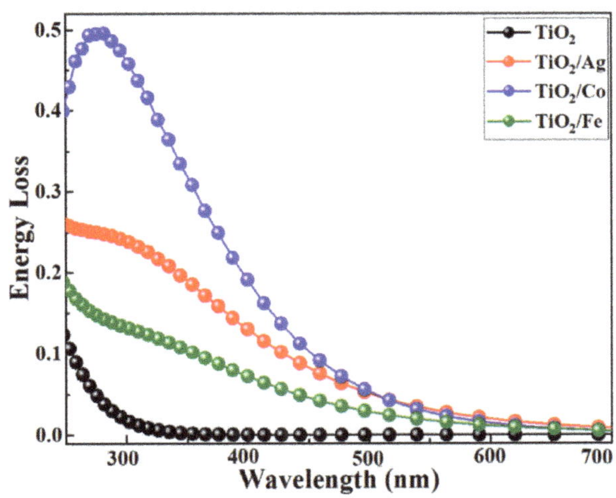

Figure 9. Energy loss vs. wavelength for pristine and transition-metal-doped TiO_2.

2.3. Elastic Properties

The crystal responses to the external parameters, such as strain and pressure, were governed by the elastic characteristics of the material, which included its mechanical stability, the modes of the phonons, and other solid-state phenomena. To understand the mechanical properties of any solid, the elastic constants (C_{ij}) listed in Table 3 were of great importance; they could be obtained by the Taylor-series expansion [52].

Table 3. Elastic constants (C_{ij}; GPa) for pure and Co-doped TiO_2 anatase comparison data.

Elastic Constants	This Work TiO_2	Other DFT [39][a] [53][b]	This Work TiO_2/Co
C_{11}	369.140	[a]399.1, [b]336.5	340.595
C_{12}	153.918	[a]167.6, [b]138.6	155.326
C_{13}	153.260	[a]159.9, [b]136.0	138.609
C_{33}	215.389	[a]250.8, [b]192.1	196.803
C_{44}	44.281	[a]70.49, [b]49.4	60.523
C_{66}	58.665	[a]61.28, [b]58.3	63.446

[a] Ref. [39]; [b] [53].

Due to the symmetry, each crystal had its own specific elastic constant. As TiO_2 (anatase) exists in tetragonal symmetry, there were six independent elastic constants: C_{11}, C_{12}, C_{13}, C_{33}, C_{44}, and C_{66}. These elastic constants correlated the stress (σ) and strain (ε) tensors using Hooke's law. Table 4 depicts the estimated values of the elastic moduli, could be derived from the Voigt–Reuss–Hill approximations and were in good agreement with existing theoretical and experimental results [54]. The mechanical stability was determined by the strain energy, which had to be positive for any elastic deformation. For tetragonal symmetry, the crystal had to fulfill the following criteria: $C_{11} > 0$, $C_{44} > 0$, $C_{66} > 0$, $C_{11} - C_{12} > 0$, and $C_{11}C_{33} + C_{12}C_{33} - 2C_{13}^2 > 0$. Our findings revealed that all the conditions were successfully satisfied, and that all the materials were mechanically stable.

Table 4. Estimated elastic moduli and allied parameters of pristine and Co-doped TiO_2.

Elastic Properties	Current Work TiO_2	Other Work TiO_2 [55]	Current Work Co dope TiO_2
B_V	208.283	221.982	193.675
B_R	192.736	204.210	176.590
B_H	200.510	213.096	185.132
G_V	62.327	51.275	66.595
G_R	55.230	32.085	63.099
G_H	58.779	41.680	64.847
E_V	170.023	142.829	179.241
E_R	151.243	91.466	169.152
E_H	160.640	117.388	174.203
υ_V	0.363	0.393	0.346
υ_R	0.369	0.425	0.340
υ_H	0.366	0.408	0.343
B_V/G_V	3.342	4.329	2.908
B_R/G_R	3.489	6.365	2.799
B_H/G_H	3.411	5.113	2.855
A^U	0.423	0.368	0.374

Bulk modulus (B) is the capability of the material volume to resist elastic deformation. It indicates the binding energy and average bond strength of atoms in a crystal [52]. Table 3 shows the bulk modulus of pristine and Co-doped TiO_2 with the values of 192.736 and 176.590 GPa, respectively, indicating a high average bond strength between Ti and O [55].

Figure 4 shows that the structural properties were attributed to the O-2p and Ti-3d states, indicating ionic and partially covalent bonds. This spatial arrangement made it arduous to modify the atomic displacement through the action of an external force. Meanwhile, the isoionic bond and covalent bond directionality made dislocation almost impossible and created obstacles to dislocation propagation for the surrounding grains. At room temperature, the pristine and doped TiO_2 were sensitive to deformation. Under the influence of an external force, if breaking occurred in the structure, it indicated brittleness, poor toughness, and a relatively low shear modulus, as shown in Table 4.

In addition, Poisson's ratio (υ) governs material brittleness and ductility, along with the expansion capability in the direction normal of the compression. A material with a Poisson's ratio less than 0.33 was considered to be fragile; otherwise, it was considered ductile [39]. Pristine TiO_2 has a higher Poisson's ratio than Co-doped TiO_2, indicating a higher ductility. Moreover, a Pugh's ratio (B/G) greater than 1.75 suggested a ductile nature, and our estimated results also indicated a ductile nature in both systems, suggesting consistency with Poisson's ratio [39].

To understand the directional dependence of physical properties, anisotropy (A^U) is very important. The null value of anisotropy indicates the isotropic nature of the material. Furthermore, it also determines the nature of the microcracks and the phase stability of crystal structures. A higher anisotropy value indicates microcracks and stress concentration. Our calculation revealed that the incorporation of Co produced fewer microcracks than found in pristine TiO_2. Moreover, the Poisson's ratio and anisotropy values were in good agreement with the literature data and are listed in Table 4. The elastic properties were related to the thermal properties, including the Grüneisen parameter and Debye temperature. The corresponding mechanical properties were promising for energy-associated applications [56].

3. Conclusions

The optoelectronic properties of pristine and doped TiO_2 were meticulously analyzed for PEC water splitting using DFT simulations. It was observed that a 2.1% incorporation of each dopant accentuated the optical absorption and the charge transport capabilities in a broad range of the solar spectrum. The calculated bandgap was reduced to 1.50 eV, 2.02 eV, and 1.58 eV in the doped variants of silver, cobalt, and iron, respectively, due to the generation of novel states and the sensitization of the materials to the visible and infrared region, indicating promising PEC water-splitting activity. Furthermore, a more negative CBM and a more positive VBM, with respect to the water oxidation and reduction, respectively, made the materials more efficacious for water splitting. In addition, the enhancement in the optical absorption due to substitutional doping increased the carrier concentration, which overcame the electron–hole pair recombination. Because of the suitable bandgap of TiO_2/Co, we estimated its elastic properties and observed that it had the highest average bond strength, as well as high ductility. Our calculations also revealed mechanically stable configurations, fewer microcracks, and lower stress concentrations in Co-doped TiO_2, as compared to pristine TiO_2. In general, substitutional doping significantly altered the optical conductivity, the dielectric function, the reflectance, and the refractive index of the materials, suggesting their technological potential in PEC water splitting.

Author Contributions: Simulation Study, A.H.; Investigation, A.H.; A.R., E.A. and M.S.K.; Writing—original draft, S.A.M.; writing—review and editing, J.J. All authors have read and agreed to the published version of the manuscript.

Funding: This work was supported by the National Research Foundation of Korea (NRF) grant funded by the Korea government (MSIT) (No. 2021R1A2C1010936).

Institutional Review Board Statement: Not applicable.

Informed Consent Statement: Not applicable.

Data Availability Statement: Not applicable.

Conflicts of Interest: The authors declare no conflict of interest.

Sample Availability: Samples of the compounds are available from the authors.

References

1. Vorontsov, A.V.; Valdés, H.; Smirniotis, P.G.; Paz, Y. Recent advancements in the understanding of the surface chemistry in TiO_2 photocatalysis. *Surfaces* **2020**, *3*, 72–92. [CrossRef]
2. Zhang, J.; Zhou, P.; Liu, J.; Yu, J. New understanding of the difference of photocatalytic activity among anatase, rutile and brookite TiO_2. *Phys. Chem. Chem. Phys.* **2014**, *16*, 20382–20386. [CrossRef]
3. Liu, X.; Li, Y.; Wei, Z.; Shi, L. A Fundamental DFT Study of Anatase (TiO_2) Doped with 3d Transition Metals for High Photocatalytic Activities. *J. Wuhan Univ. Technol. Sci. Ed.* **2018**, *33*, 403–408. [CrossRef]
4. Ilyas, M.; Hussain, A.; Rauf, A.; Rahman, I.U.; Naveed, A.; Basit, M.A.; Rabbani, F.; Khan, S.U.; Ahmed, E.; Hussain, M.; et al. Tailoring the antifouling agent titanium dioxide in the visible range of solar spectrum for photoelectrochemical activity with hybrid DFT & DFT+ U approach. *Mater. Today Commun.* **2021**, *27*, 102366.
5. Dharmale, N.; Chaudhury, S.; Kar, J. Technology. Pressure-Induced Phase Transition Study on Brookite to Rutile TiO_2 Transformation. *ECS J. Solid State Sci. Technol.* **2021**, *10*, 071021. [CrossRef]
6. Chen, T.; Chen, W.-L.; Foley, B.J.; Lee, J.; Ruff, J.P.; Ko, J.P.; Brown, C.M.; Harriger, L.W.; Zhang, D.; Park, C.; et al. Origin of long lifetime of band-edge charge carriers in organic–inorganic lead iodide perovskites. *Proc. Natl. Acad. Sci. USA* **2017**, *114*, 7519–7524. [CrossRef] [PubMed]
7. Huy, H.A.; Aradi, B.; Frauenheim, T.; Deák, P. Calculation of carrier-concentration-dependent effective mass in Nb-doped anatase crystals of TiO_2. *Phys. Rev. B* **2011**, *83*, 155201. [CrossRef]
8. Tamirat, A.G.; Rick, J.; Dubale, A.A.; Su, W.-N.; Hwang, B.-J. Using hematite for photoelectrochemical water splitting: A review of current progress and challenges. *Nanoscale Horiz.* **2016**, *1*, 243–267. [CrossRef]
9. Du, C.; Yang, X.; Mayer, M.T.; Hoyt, H.; Xie, J.; McMahon, G.; Bischoping, G.; Wang, D. Hematite-based water splitting with low turn-on voltages. *Angew. Chem.* **2013**, *125*, 12924–12927. [CrossRef]
10. Sivula, K.; Le Formal, F.; Grätzel, M. Solar water splitting: Progress using hematite (α-Fe_2O_3) photoelectrodes. *Chemsuschem* **2011**, *4*, 432–449. [CrossRef]
11. Wang, Y.; Tian, W.; Chen, C.; Xu, W.; Li, L. Tungsten trioxide nanostructures for photoelectrochemical water splitting: Material engineering and charge carrier dynamic manipulation. *Adv. Funct. Mater.* **2019**, *29*, 1809036. [CrossRef]
12. Enesca, A.; Duta, A.; Schoonman, J. Study of photoactivity of tungsten trioxide (WO3) for water splitting. *Thin Solid Films* **2007**, *515*, 6371–6374. [CrossRef]
13. Park, J.H.; Park, O.O.; Kim, S. Photoelectrochemical water splitting at titanium dioxide nanotubes coated with tungsten trioxide. *Appl. Phys. Lett.* **2006**, *89*, 163106. [CrossRef]
14. Fujishima, A.; Zhang, X.; Tryk, D.A. TiO_2 photocatalysis and related surface phenomena. *Surf. Sci. Rep.* **2008**, *63*, 515–582. [CrossRef]
15. Hisatomi, T.; Kubota, J.; Domen, K. Recent advances in semiconductors for photocatalytic and photoelectrochemical water splitting. *Chem. Soc. Rev.* **2014**, *43*, 7520–7535. [CrossRef]
16. Khan, S.U.; Al-Shahry, M.; Ingler, W.B., Jr. Efficient photochemical water splitting by a chemically modified n-TiO_2. *Science* **2002**, *297*, 2243–2245. [CrossRef]
17. Guerrero, A.; Bisquert, J. Perovskite semiconductors for photoelectrochemical water splitting applications. *Curr. Opin. Electrochem.* **2017**, *2*, 144–147. [CrossRef]
18. Wang, C.; Qiu, H.; Inoue, T.; Yao, Q. Band gap engineering of $SrTiO_3$ for water splitting under visible light irradiation. *Int. J. Hydrogen Energy* **2014**, *39*, 12507–12514. [CrossRef]
19. Yang, Y.; Niu, S.; Han, D.; Liu, T.; Wang, G.; Li, Y. Progress in developing metal oxide nanomaterials for photoelectrochemical water splitting. *Adv. Energy Mater.* **2017**, *7*, 1700555. [CrossRef]
20. Chandrasekaran, S.; Yao, L.; Deng, L.; Bowen, C.; Zhang, Y.; Chen, S.; Lin, Z.; Peng, F.; Zhang, P. Recent advances in metal sulfides: From controlled fabrication to electrocatalytic, photocatalytic and photoelectrochemical water splitting and beyond. *Chem. Soc. Rev.* **2019**, *48*, 4178–4280. [CrossRef]
21. Rueda, D.; Arias, V.; Zhang, Y.; Cabot, A.; Agudelo, A.C.; Cadavid, D. Monitoring; Management. Low-cost tangerine peel waste mediated production of titanium dioxide nanocrystals: Synthesis and characterization. *Environ. Nanotechnol. Monit. Manag.* **2020**, *13*, 100285.
22. Kanakaraju, D.; Glass, B.D.; Oelgemöller, M. Titanium dioxide photocatalysis for pharmaceutical wastewater treatment. *Environ. Chem. Lett.* **2014**, *12*, 27–47. [CrossRef]
23. Yang, X.; Cui, F.; Guo, X.; Li, D. Effects of nanosized titanium dioxide on the physicochemical stability of activated sludge flocs using the thermodynamic approach and Kelvin probe force microscopy. *Water Res.* **2013**, *47*, 3947–3958. [CrossRef]
24. Pelaez, M.; Nolan, N.T.; Pillai, S.C.; Seery, M.K.; Falaras, P.; Kontos, A.G.; Dunlop, P.S.; Hamilton, J.W.; Byrne, J.A.; O'shea, K.; et al. A review on the visible light active titanium dioxide photocatalysts for environmental applications. *Appl. Catal. B Environ.* **2012**, *125*, 331–349. [CrossRef]
25. Chen, S.; Takata, T.; Domen, K. Particulate photocatalysts for overall water splitting. *Nat. Rev. Mater.* **2017**, *2*, 1–17. [CrossRef]

26. Qian, R.; Zong, H.; Schneider, J.; Zhou, G.; Zhao, T.; Li, Y.; Yang, J.; Bahnemann, D.W.; Pan, J.H. Charge carrier trapping, recombination and transfer during TiO_2 photocatalysis: An overview. *Catal. Today* **2019**, *335*, 78–90. [CrossRef]
27. Zhang, Y.C.; Yang, M.; Zhang, G.; Dionysiou, D.D. HNO_3-involved one-step low temperature solvothermal synthesis of N-doped TiO_2 nanocrystals for efficient photocatalytic reduction of Cr (VI) in water. *Appl. Catal. B Environ.* **2013**, *142*, 249–258. [CrossRef]
28. Zhang, Y.C.; Li, J.; Xu, H.Y. One-step in situ solvothermal synthesis of SnS_2/TiO_2 nanocomposites with high performance in visible light-driven photocatalytic reduction of aqueous Cr (VI). *Appl. Catal. B Environ.* **2012**, *123*, 18–26. [CrossRef]
29. Basavarajappa, P.S.; Patil, S.B.; Ganganagappa, N.; Reddy, K.R.; Raghu, A.V.; Reddy, C.V. Recent progress in metal-doped TiO_2, non-metal doped/codoped TiO_2 and TiO_2 nanostructured hybrids for enhanced photocatalysis. *Int. J. Hydrogen Energy* **2020**, *45*, 7764–7778. [CrossRef]
30. Yalçın, Y.; Kılıç, M.; Çınar, Z. The role of non-metal doping in TiO_2 photocatalysis. *J. Adv. Oxid. Technol.* **2010**, *13*, 281–296. [CrossRef]
31. Wang, Y.; Zhang, R.; Li, J.; Li, L.; Lin, S. First-principles study on transition metal-doped anatase TiO_2. *Nanoscale Res. Lett.* **2014**, *9*, 46. [CrossRef] [PubMed]
32. He, G.; Zhu, L.; Liu, M.; Fang, Q.; Zhang, L. Optical and electrical properties of plasma-oxidation derived HfO_2 gate dielectric films. *Appl. Surf. Sci.* **2007**, *253*, 3413–3418. [CrossRef]
33. Pandey, S.C.; Xu, X.; Williamson, I.; Nelson, E.B.; Li, L. Electronic and vibrational properties of transition metal-oxides: Comparison of GGA, GGA+ U, and hybrid approaches. *Chem. Phys. Lett.* **2017**, *669*, 1–8. [CrossRef]
34. Orhan, O.K.; O'Regan, D.D. TDDFT+ U: A critical assessment of the Hubbard U correction to exchange-correlation kernels and potentials. *Phys. Rev. B* **2019**, *99*, 165120. [CrossRef]
35. Parr, R.G. Density functional theory of atoms and molecules. In *Horizons of Quantum Chemistry*; Springer: Dordrecht, The Netherlands, 1980; pp. 5–15.
36. Arshad Javid, M.; Khan, Z.U.; Mehmood, Z.; Nabi, A.; Hussain, F.; Imran, M.; Nadeem, M.; Anjum, N. Structural, electronic and optical properties of $LiNbO_3$ using GGA-PBE and TB-mBJ functionals: A DFT study. *Int. J. Mod. Phys. B* **2018**, *32*, 1850168. [CrossRef]
37. Nisar, J.; Århammar, C.; Jämstorp, E.; Ahuja, R. Optical gap and native point defects in kaolinite studied by the GGA-PBE, HSE functional, and GW approaches. *Phys. Rev. B* **2011**, *84*, 075120. [CrossRef]
38. Sharma, S.B. Optoelectronic and elastic response of fluorinated hexagonal boron nitride monolayer. *ChemRxiv* **2021**. *ahead of print*.
39. Mazumder, J.T.; Mayengbam, R.; Tripathy, S. Theoretical investigation on structural, electronic, optical and elastic properties of TiO_2, SnO_2, ZrO_2 and HfO_2 using SCAN meta-GGA functional: A DFT study. *Mater. Chem. Phys.* **2020**, *254*, 123474. [CrossRef]
40. Uspenskii, Y.A.; Kulatov, E.; Halilov, S. Effect of anisotropy on magneto-optical properties of uniaxial crystals: Application to CrO_2. *Phys. Rev. B* **1996**, *54*, 474. [CrossRef]
41. Wanbayor, R.; Ruangpornvisuti, V. Adsorption of CO, H_2, N_2O, NH_3 and CH_4 on the anatase TiO_2 (0 0 1) and (1 0 1) surfaces and their competitive adsorption predicted by periodic DFT calculations. *Mater. Chem. Phys.* **2010**, *124*, 720–725. [CrossRef]
42. Howard, C.; Sabine, T.; Dickson, F. Structural and thermal parameters for rutile and anatase. *Acta Crystallogr. Sect. B Struct. Sci.* **1991**, *47*, 462–468. [CrossRef]
43. Pan, Y.; Wen, M. Noble metals enhanced catalytic activity of anatase TiO_2 for hydrogen evolution reaction. *Int. J. Hydrogen Energy* **2018**, *43*, 22055–22063. [CrossRef]
44. Wu, H.-C.; Li, S.-H.; Lin, S.-W. Effect of Fe concentration on Fe-doped anatase TiO_2 from GGA+ U calculations. *Int. J. Photoenergy* **2012**, *2012*, 823498. [CrossRef]
45. Pan, L.; Ai, M.; Huang, C.; Yin, L.; Liu, X.; Zhang, R.; Wang, S.; Jiang, Z.; Zhang, X.; Zou, J.-J.; et al. Manipulating spin polarization of titanium dioxide for efficient photocatalysis. *Nat. Commun.* **2020**, *11*, 418. [CrossRef] [PubMed]
46. Cendula, P.; Tilley, S.D.; Gimenez, S.; Bisquert, J.; Schmid, M.; Grätzel, M.; Schumacher, J.O. Calculation of the energy band diagram of a photoelectrochemical water splitting cell. *J. Phys. Chem. C* **2014**, *118*, 29599–29607. [CrossRef]
47. Chen, S.; Wang, L.-W. Thermodynamic oxidation and reduction potentials of photocatalytic semiconductors in aqueous solution. *Chem. Mater.* **2012**, *24*, 3659–3666. [CrossRef]
48. Lin, Y.; Jiang, Z.; Zhu, C.; Hu, X.; Zhu, H.; Zhang, X.; Fan, J.; Lin, S.H. The optical absorption and hydrogen production by water splitting of (Si, Fe)-codoped anatase TiO_2 photocatalyst. *Int. J. Hydrog. Energy* **2013**, *38*, 5209–5214. [CrossRef]
49. Dash, L.; Vast, N.; Baranek, P.; Cheynet, M.-C.; Reining, L. Electronic structure and electron energy-loss spectroscopy of ZrO_2 zirconia. *Phys. Rev. B* **2004**, *70*, 245116. [CrossRef]
50. Sharma, P.; Katyal, S. Determination of optical parameters of a-$(As_2Se_3)_{90}Ge_{10}$ thin film. *J. Phys. D Appl. Phys.* **2007**, *40*, 2115. [CrossRef]
51. Khoshman, J.M.; Kordesch, M.E. Optical properties of a-HfO_2 thin films. *Surf. Coat. Technol.* **2006**, *201*, 3530–3535. [CrossRef]
52. Pfrommer, B.G.; Côté, M.; Louie, S.G.; Cohen, M.L. Relaxation of crystals with the quasi-Newton method. *J. Comput. Phys.* **1997**, *131*, 233–240. [CrossRef]
53. Yao, H.; Ouyang, L.; Ching, W.Y. Ab initio calculation of elastic constants of ceramic crystals. *J. Am. Ceram. Soc.* **2007**, *90*, 3194–3204. [CrossRef]
54. Mahmood, T.; Malik, H.; Batool, R.; Perveen, Z.; Saleemi, F.; Rasheed, H.; Saeed, M.; Cao, C.; Rizwan, M. Elastic, electronic and optical properties of anatase TiO_2 under pressure: A DFT approach. *Chin. J. Phys.* **2017**, *55*, 1252–1263. [CrossRef]

55. Liu, X.; Fu, J. Electronic and elastic properties of the tetragonal anatase TiO_2 structure from first principle calculation. *Optik* **2020**, *206*, 164342. [CrossRef]
56. Lü, X.; Yang, W.; Quan, Z.; Lin, T.; Bai, L.; Wang, L.; Huang, F.; Zhao, Y. Enhanced electron transport in Nb-doped TiO_2 nanoparticles via pressure-induced phase transitions. *J. Am. Chem. Soc.* **2014**, *136*, 419–426. [CrossRef] [PubMed]

molecules

MDPI

Article

The Stability of a Mixed-Phase Barium Cerium Iron Oxide under Reducing Conditions in the Presence of Hydrogen

Benjamin Rosen and Karl Sohlberg *

Chemistry Department, College of Arts and Sciences, Drexel University, Philadelphia, PA 19104, USA
* Correspondence: karl.william.sohlberg@drexel.edu

Abstract: Metal oxide perovskite materials show promise for use as hydrogen separation membranes, but metal oxides can dehydrate in the presence of hydrogen to the point of decomposition. The stability of a material in the presence of hydrogen is necessary for an effective hydrogen separation membrane. The stability of a mixed phase metal oxide perovskite ($BaCe_{0.85}Fe_{0.15}O_{3-\delta}$-$BaCe_{0.15}Fe_{0.85}O_{3-\delta}$) was investigated using first-principles thermodynamics calculations based on density functional theory to examine the possible reduction processes on the surface of the material. It was found that for either phase of the material, the loss of H_2 becomes thermodynamically favorable over the formation of oxygen vacancies once oxygen vacancy defects exist on the surface. Additionally, both phases of the material become more stable with respect to the dehydration or loss of oxygen with increasing concentrations of surface oxygen vacancies. Under the conditions of commercial hydrogen production (~400–1100 K), it is more thermodynamically favorable for H_2 to desorb from the $BaCe_{0.85}Fe_{0.15}O_{3-\delta}$ phase. Examination of the atomic-scale structure indicates that the degree of coordination of surface metal atoms in this material may control the stability of the material in reducing environments.

Keywords: hydrogen; stability; DFT; membrane; perovskite oxide; separation

1. Introduction

As humanity moves away from fossil fuels as a primary energy source, hydrogen is widely seen as a clean alternative energy carrier. Consequently, hydrogen demand is predicted to outpace production over the next decade [1,2]. A fundamental bottleneck in current commercial methods of hydrogen production, steam reforming or dehydrogenation of hydrocarbons, is the energy- and time-intensive process of hydrogen isolation from the resulting product mixtures [3]. Membrane reactor systems are an attractive solution to remediate this process bottleneck. Membrane reactors have not yet seen widespread commercial adoption because current metallic membrane materials are prohibitively costly and have short cycle lives [4,5]. Metal oxide ceramics that can function as hydrogen separation membranes are a class of materials that have the potential to address the issues with current commercially viable hydrogen separation membranes [6,7]. Many metal oxide materials decompose via dehydration in hydrogen rich environments—a consequence of reduction by hydrogen resulting in the dehydration of the material [8,9]. For a material to properly function as a hydrogen separation membrane, the material must be stable with respect to both reduction and oxidation in hydrogen-rich environments so that hydrogen gas can repeatedly adsorb at the retentate surface and desorb from the permeate surface without damaging the membrane. Additionally, for hydrogen gas to be collected from the permeate surface of a hydrogen separation membrane, desorption of hydrogen gas must be thermochemically preferred over dehydration. Dehydration here only refers to the evolution of water as a product, not the loss of absorbed water. Structurally stable materials are paramount for the development and manufacture of commercial hydrogen separation membranes. In order to design improved ceramic oxide hydrogen separation membranes, it is necessary to ensure the materials will be stable in the harsh chemical environments of commercial hydrogen production.

Some metal oxide perovskite materials appear to be stable under desorption of hydrogen over a wide temperature range, and exhibit hydrogen flux rates comparable to metal membranes [6,10]. Stability combined with unique optical, electronic, and catalytic properties gives materials with the perovskite structure wide-ranging applications [11]. Cheng et al. [10] reported an investigation of a ternary metal oxide ceramic as a hydrogen separation membrane material. The material investigated by Cheng et al. [10] was reported to be a mixed-phase material composed of two complimentary defect perovskite structures, $BaCe_{0.85}Fe_{0.15}O_{3-\delta}$ and $BaCe_{0.15}Fe_{0.85}O_{3-\delta}$ (BCF8515 and BCF1585, respectively), in a 50/50 stoichiometric ratio. Cheng et al. [10] did not report the detection of water in the permeate gas sweep. Only H_2, N_2 bleed over from the retentate side of the membrane, and the argon permeate sweep gas were reported [10]. Similar results were reported by Xia et al. [12]. Many metal oxides reduce and desorb water in the presence of H_2 due to the thermodynamic stability of water [8,13,14]. The apparent lack of dehydration at the permeate surface upon reduction by H_2 indicates that BCF8515-BCF1585 may be stable with respect to reduction in hydrogen atmospheres. Stability in the presence of hydrogen is necessary for commercially viable materials. Understanding what material properties lead to stability in reducing hydrogen atmospheres is fundamental to designing improved hydrogen separation membrane materials.

A theoretical investigation of the stability of BCF8515-BCF1585 to reduction at the permeate surface was performed. To do so, the thermodynamics of several possible competing and sequential reactions at the surface of BCF8515 and BCF1585 crystals were determined from first principles with density functional theory (DFT) calculations, the results and interpretation of which are reported here.

2. Methods

2.1. Computational Details

2.1.1. Software Parameters

All electronic structure calculations were performed with density functional theory using the PW91 [15] approximation to the exchange-correlation and were carried out with the Vienna ab initio Simulation Package (VASP) [16–18]. All calculations employed a plane-wave basis subject to a cutoff energy of 400 eV. The pseudopotentials used were all the standard VASP-provided projector augmented wave (PAW) pseudopotentials for use with general gradient approximation (GGA) exchange-correlation functionals [19]. All geometry optimizations were performed using either the conjugate gradient relaxation algorithm, or the RMM-DIIS relaxation algorithm [16]. The RMM-DIIS algorithm was only utilized in cases where the conjugate gradient algorithm resulted in a bracketing error due to the nuclear geometry being very near a minimum energy configuration. All structural relaxations were terminated when the change in the forces was less than 0.02 $\frac{eV}{Å}$. The k-point sampling density for all calculations was determined by VASP's built-in automatic k-point grid selector. All k-point grids utilized can be found in S1–S4 of the Supplementary Materials. Gaussian smearing with a smearing width of 0.2 eV was used for all calculations. The 0K total electronic energy values were utilized for all free energy calculations in the standard approximation [20,21].

2.1.2. Structural Models

The initial bulk unit cell structures for all three materials investigated were built by modifying an undoped perovskite material ($BaCeO_3$ or $BaFeO_3$) having the desired symmetry (cubic or orthorhombic). A supercell of the undoped material was built, such that a B-site metal cation could be substituted for a dopant atom to yield the desired stoichiometry. All bulk structures can be found in S1 of the SI. All *dopant* defects were substitutional only. All structures had 1 dopant atom per 7 atoms of the element being substituted. This yielded slab stoichiometries of $Ba_{16}Ce_{14}Fe_2O_{48}$ (to represent BCF8515) and $Ba_{16}Ce_2Fe_{14}O_{48}$ (to represent BCF1585) for materials containing no oxygen vacancy defects. Oxygen vacancy defects were added to the surface of each material sequentially by removing a single oxygen atom per surface supercell at each stage of decomposition. All oxygen vacancies were added on the surface of the respective slab models. The ratios of Ce to Fe used were selected to achieve the closest ratio to 15% dopant (15% by atom count, not mass percent) with unit cells sufficiently small to remain computationally tractable. For all structures investigated, all dopant defects were located within the bulk of the material or on the bottom exposed surface, not the surface interacting with adsorbed species and containing oxygen defects.

The preferred surface exposure for each material investigated was determined by finding the surface structure that minimized the surface energy. Each surface energy calculation was obtained by performing a geometry relaxation on a slab with the desired Miller index surface exposed and calculating the surface energy by

$$\text{Surface Energy} = \frac{E_{slab} - n * E_{bulk}}{2 * A_{slab}} \tag{1}$$

where E_{slab} is the total energy of the slab, E_{bulk} is the total energy of a unit cell of the bulk material, n is a scaling factor to adjust the total energy of the bulk to the number of unit cell equivalents in the slab, and A_{slab} is the surface area of the exposed surface. Every slab utilized for surface energy calculations was cleaved from 2 bulk unit cells (n = 2). The bottom two layers of atoms were frozen for each surface energy calculation to simulate the bulk material that exists between two exposed surfaces. In asymmetric cases, a dipole correction was applied along the vector normal to the exposed surface. For all slab calculations symmetric slabs were utilized wherever possible. The surface energies were calculated for the lowest and second-lowest order Miller index surfaces for each material. The lowest energy surfaces identified for each material were (010) surface for cubic BCF8515, (101) for cubic BCF1585, and (011) surface for orthorhombic BCF8515. All stability calculations were performed on the respective lowest energy surface for each material. All slab structures can be found in S2–S4 of the SI. Zhu et al. [13] reported that the structure of BCF1585 remains in the cubic space group for temperatures below 1200 °C. Since the cubic structure of BCF1585 presented by Cheng et al. [10] is unambiguous, no investigation of an orthorhombic phase of BCF1585 was performed.

2.1.3. Selection of Crystal Structures

Selection of the crystal structure for BCF1585 was accomplished by comparing the crystal structure presented by Cheng et al. [10] with the common cubic space groups of $BaFeO_3$. BCF1585 was identified as being derived from $BaFeO_3$ in the cubic $pm\bar{3}m$ space group.

For the BCF8515 phase, Zhu et al. [13], Cheng et al. [10], and Xia et al. [12] all reported an orthorhombic structure; however, none reported a possible space group. Cheng et al. [10] reported BCF8515 as having an orthorhombic symmetry that results from an elongation of one axis from the $BaCeO_3$ $pm\bar{3}m$ cubic structure. For $BaCeO_3$, elongation of one axis results in an orthorhombic structure having symmetry in the *pnma* space group. Cheng et al. [10] reported that both phases (Fe or Ce dominant) form structures with the same symmetry as the corresponding undoped materials ($BaFeO_3$ and $BaCeO_3$, respectively). As per Knight [22], $BaCeO_3$ undergoes 3 temperature-dependent phase transitions. According to Knight [22], $BaCeO_3$ transitions to the cubic $pm\bar{3}m$ space group at 900 °C. Cheng et al. [10] reported calcining both BCF8515-BCF1585 samples at 1370 °C. The subsequent hydrogen permeation testing was performed from 850 to 950 °C. The temperatures at which the BCF8515–BCF1585 samples were treated and tested were high enough to facilitate a transformation to the cubic phase, assuming BCF8515 behaves similarly to $BaCeO_3$.

In an effort to determine the most probable structure of BCF8515, three possible structures were compared: two orthorhombic structures derived by doping the *pnma* phase of $BaCeO_3$ with an iron defect (Fe substitution at each of the two possible Ce sites) and a cubic phase derived from the $pm\bar{3}m$ phase of $BaCeO_3$. Each bulk structure was relaxed as per the method described in Sections 2.1.1 and 2.1.2 for unit cells of the same stoichiometry. The lower energy of the two orthorhombic phases was identified from total energy calculations and selected for stability investigation. The cubic structure had the highest total energy. In an attempt to resolve the ambiguity in the results presented by Cheng et al. [10], and the contradiction with the preferred phase under the reaction conditions, based on the results of Knight [22], the predicted 2θ plots for the orthorhombic and cubic structures were compared. The calculated 2θ plots can be found in Figure 1. From the 2θ plots, it can be seen that the only noticeable peak shifts that occur between the two structures can be found at 2θ > 80°. The orthorhombic structure does have some peak broadening compared to the cubic structure; however, this feature is not pronounced until 2θ > 80°. Since the ambiguity between the lowest energy structure and the structure likely to exist under the reaction conditions could not be resolved, both the orthorhombic and cubic phases of BCF8515 were investigated, and the results for both are presented herein.

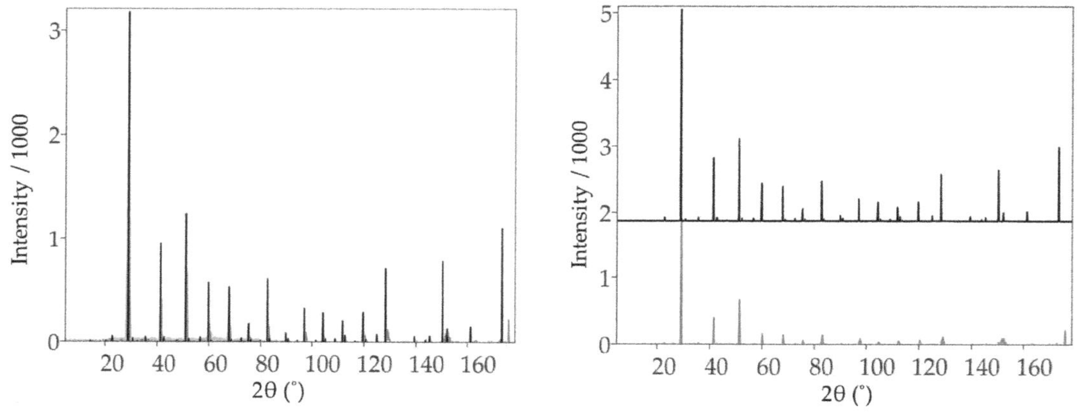

(a) (b)

Figure 1. (**a**) Simulated X-ray diffraction patterns for orthorhombic (pink) and cubic (black) BCF8515 for 2θ from 0° to 180°, overlayed with data extracted from Zhu et al. [13] (green). (**b**) Simulated X-ray diffraction patterns for orthorhombic and cubic BCF8515 for 2θ from 0° to 180°, stacked. Simulated data generated using CrystalDiffract®, CrystalMaker Software Ltd. (Oxford, UK), (www.crystalmaker.com accessed on 26 December 2022).

2.2. Stability Study Model

2.2.1. Protonation: Decomposition or Desorption

The goal of this study was to examine the thermodynamics of forming reduction products from the protonated surface of barium cerium iron oxides to investigate the stability of these materials in the chemical environment of a hydrogen-generating membrane reactor. Three possible reactions yielding gas-phase products were considered: 1. The desorption of H_2. 2. The desorption of H_2O and the formation of an oxygen vacancy. 3. The spontaneous formation of an oxygen vacancy. These reactions are represented by reactions (I)–(III) below, where S denotes the slab model and V^O an oxygen vacancy.

$$2H/S \rightarrow H_2 + S\,(I)$$
$$2H/S \rightarrow H_2O + V^O/S\,(II)$$
$$2H/S \rightarrow \frac{1}{2}O_2 + 2HV^O/S\,(III)$$

The enthalpy and entropy changes were calculated for each of the reactions (I)–(III) and were used to predict the temperature at which the Gibbs free energy for each process becomes negative (passes equilibrium to become spontaneous). This was repeated for surfaces with an increasing number of oxygen vacancies for each material. The thermodynamically favored processes for each possible reaction on each possible surface were determined by calculating the temperature-dependent Gibbs free energy change for each reaction and comparing across the reactions to determine the thermodynamically preferential reaction on each surface.

2.2.2. Enthalpy Change Calculations

The total change in enthalpy for each representative reaction was calculated by first performing an electronic total energy calculation on each reactant and product. The change in enthalpy was calculated as

$$\Delta\text{Enthalpy} = \Sigma \text{ total energy products} - \Sigma \text{ total energy reactants} \tag{2}$$

since PV work is negligible for the systems of interest due to the low pressure (380 pascal) of products at the surface. The total energy calculations for each gas-phase molecule (H_2, O_2, H_2O) were performed by placing a single molecule in the center of a 10 × 10 × 10 Å unit cell and performing a full geometry optimization of the molecule. For all calculations that utilize $\frac{1}{2}$ O_2, the total enthalpy of O_2 was multiplied by a factor of $\frac{1}{2}$.

2.2.3. Entropy Change Calculations

Calculation of the change in entropy from reactants to products in the representative reactions began with the following assumptions: All the "reactant" species for all the representative reactions were bound to periodic solids, so there could only be vibrational degrees of freedom (movement of atoms in the desorbing species relative to the surface slab, i.e., no rotation or translation). The products in the representative reactions included gas-phase species, which have vibrational, rotational, and translational degrees of freedom. All reactions were assumed to be in the electronic ground state, so electronic degrees of freedom were not considered in the canonical ensemble. The change in entropy was taken as the entropy difference between the vibrational degrees of freedom for the atoms constituting the gas-phase species when bound to the solid surface and the total entropy of the gas-phase species. The entropy change of the remainder of the surface was neglected because it was assumed to be an insignificant change in relation to the entropy change due to the formation of product gas [23]. This yielded a change in entropy for the formation of a gas-phase molecule from the surface of interest.

The absolute entropies were calculated in the standard way,

$$S = k_B \ln(Q) \tag{3}$$

where k_B is Boltzmann's constant and Q is the canonical partition function representing all the accessible states. For the gas-phase species, the canonical partition function was taken as the product of the q_{rot} and q_{vib} (rotation and vibration, respectively) partition functions under the rigid rotor and harmonic oscillator approximations,

$$Q_{gas} = q_{vib} \times q_{rot} \tag{4}$$

since the electronic partition function was assumed to be in the ground state ($q_{electronic} = 1$). The vibrational and rotational partition functions were calculated in the standard way:

$$q_{vib} = \prod_{i=1}^{n} \frac{\exp\left(-\frac{h\nu_i}{k_B T}\right)}{1 - \exp\left(-\frac{h\nu_i}{k_B T}\right)} \tag{5}$$

where ν_i is the ith vibrational frequency, k_B is Boltzmann's constant, h is Planck's constant, T is the temperature in kelvin, and

$$q_{rot} = \frac{\pi^{\frac{1}{2}}}{\sigma} \prod_{i=1}^{n} \left(\frac{8\pi^2 I_i k_B T}{h^2} \right)^{\frac{1}{2}} \quad (6)$$

where I_i is the ith rotational moment of inertia, k_B is Boltzmann's constant, h is Planck's constant, σ is a factor correcting for degenerate rotations, and T is the temperature in kelvin. All vibrational frequencies were obtained by performing a vibrational eigenvalue calculation restricted to the atoms of interest with VASP. All moments of inertia were calculated from the equilibrium nuclear coordinates as calculated with VASP. The translational entropy was approximated using the Sakur–Tetrode equation for translational entropy [24], and was added to the total entropy of the gas-phase species, in the same manner as Cai and Sohlberg [25].

$$S_{trans} = k_B \cdot \left(\ln\left(\frac{k_B T}{P} \right) + \frac{3}{2} \cdot \ln\left(\frac{2\pi m k_B T}{h^2} \right) + \frac{5}{2} \right) \quad (7)$$

where k_B is Boltzmann's constant, T is the temperature in Kelvin, P is the partial pressure of the gas in pascals, and m is the mass of the gas.

For the solid-phase reactants, Q was taken as just the vibrational partition function for the atoms that become gas-phase species in the specific reaction under investigation. The total entropy change was calculated as the difference in the entropy of the reactants and products:

$$\Delta S_{total} = k_B \ln\left(\frac{Q_{gas}}{Q_{solid}} \right) + S_{trans} \quad (8)$$

where Q_{gas} is the canonical partition function for the product gas, Q_{solid} is the partition function for the reactants on the surface.

The entropy change for each reaction was calculated at a pressure of 380 pascal, the permeate pressure as reported by Cheng et al. [10], while allowing temperature to remain an independent variable. The temperature-dependent entropy change was substituted into the equation for the Gibbs free energy. The temperature-dependent Gibbs free energy change for each reaction was plotted on the same set of axes for each respective surface and compared. From the changes in Gibbs free energy, the most thermodynamically favorable reactions were identified for each surface. The process was continued stepwise by sequentially increasing numbers of oxygen vacancies until the free energy change for the formation of an oxygen vacancy was greater than that for the desorption of H_2 over all relevant temperatures.

3. Results

3.1. Results of Free Energy Change Analysis

The method prescribed in Section 2 (Methods) was applied to the three structures of interest: orthorhombic BCF8515, cubic BCF8515, and cubic BCF1585. Reactions (I)–(III) (see Section 2.2.1) were modeled on the lowest energy surface for each respective material, and each relevant oxygen-vacancy-containing derivative structure. The change in free energy was compared across reactions (I)–(III). For each of reactions (I)–(III) from each surface examined, the enthalpy change was found to be positive, demonstrating that all three reactions are endothermic. The entropy change was found to be positive over all temperature ranges for all three reactions considered. Thus, at elevated temperatures, the reactions reach equilibrium ($\Delta G = 0$), and eventually became spontaneous ($\Delta G < 0$). At relevant temperatures, the relative free energy changes were found to be dominated by the enthalpy changes.

3.1.1. Orthorhombic BCF8515

Cheng et al. [10] reported that the BCF8515 they synthesized had an orthorhombic structure. The orthorhombic structure utilized in this work was found to be the lowest energy structure of BCF8515. A symmetric slab was cleaved, exposing the (011) surface, as discussed in Section 2.1.2. The free energy change, ΔG, was calculated for reactions (I)–(III) occurring from the (011) surface of orthorhombic BCF8515 with zero, one, and two oxygen vacancy defects, respectively. ΔG vs. temperature for each of these nine cases can be found in Figure 2. All the equations of the lines plotted in Figure 2 can be found in S5 of the SI.

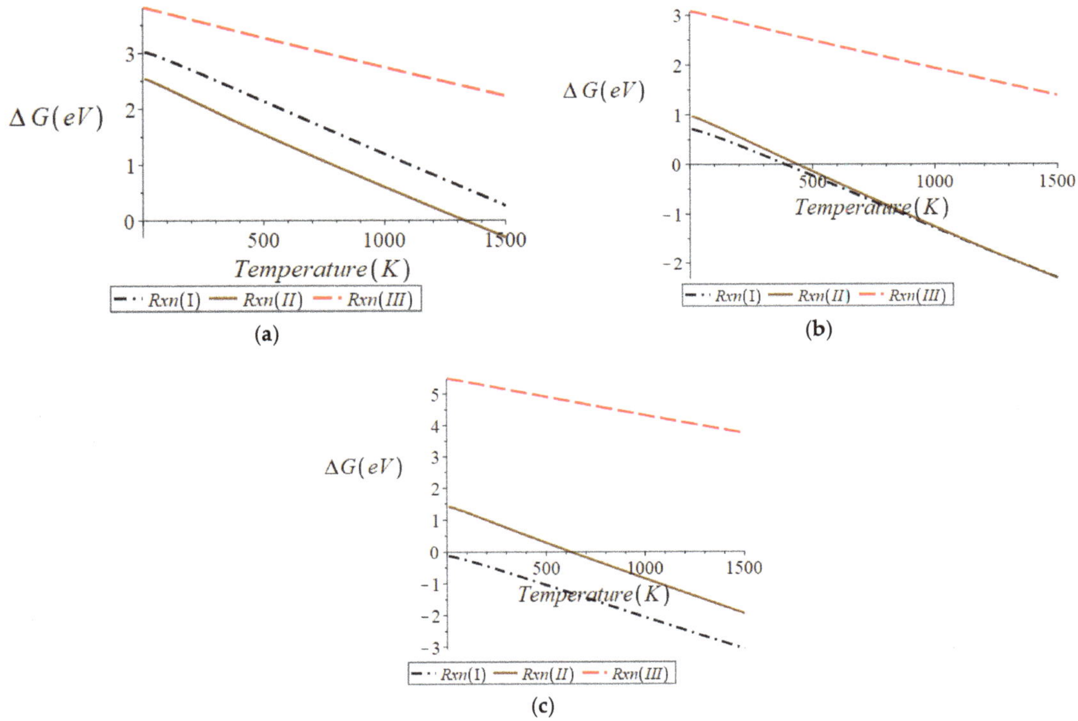

Figure 2. Surface of orthorhombic BCF8515 for (**a**) oxygen-vacancy-free surface, (**b**) surface with 1 oxygen vacancy, (**c**) surface with 2 oxygen vacancies. Plots were generated in Maple^{TM}.

From Figure 2, it can be seen that starting from the (011) surface containing no oxygen vacancies ("reactant" slab stoichiometry $Ba_{16}Ce_{14}Fe_2H_2O_{48}$), the smallest free energy change over all temperatures is that of reaction (II) (formation of $H_2O_{(g)}$). Thus, reaction (II) is the most thermodynamically favorable reaction for a surface with no oxygen vacancy defects. Since reaction (II) generates an oxygen vacancy, the free energy changes for reactions (I)–(III) were calculated for a surface containing one oxygen vacancy ("reactant" slab stoichiometry $Ba_{16}Ce_{14}Fe_2H_2O_{47}$). From Figure 2b it can be seen that the formation of H_2 from the surface with one oxygen vacancy becomes spontaneous (ΔG < 0) at the lowest temperature, though at high temperatures the free energy change for reaction (II) crosses that for reaction (I) (approximately 1200 K). For this reason, reactions (I)–(III) were investigated for the surface containing two oxygen vacancies ("reactant" slab stoichiometry $Ba_{16}Ce_{14}Fe_2H_2O_{46}$). From Figure 2c it can be seen that again reaction (I), formation of H_2, is more thermodynamically favorable than reactions (II) or (III) and is spontaneous over all temperature ranges. These results demonstrate that the surface of orthorhombic BCF8515 becomes stable to reduction once 0.0215 oxygen defects per Å^2 exist on the surface of the material. (This value and other surface oxygen vacancy threshold concentrations reported here are upper limits. The actual threshold may be lower. Investigating lower vacancy concentrations would involve calculations employing computationally demanding larger supercells).

In summary, these results indicate that if orthorhombic BCF8515 exists with surface oxygen vacancy defects at a concentration exceeding 0.0215 oxygen vacancy defects per $Å^2$, the loss of H_2 from the protonated surface becomes the dominant reduction process. It was found that the spontaneous formation of oxygen vacancies via a loss of $\frac{1}{2}$ O_2 does not occur, except at physically irrelevant temperatures. This indicates that all oxygen vacancies in the material will form either intrinsically upon crystallization, or via the dehydration of a protonated surface.

3.1.2. Cubic BCF8515

Cubic BCF8515 was examined in the same manner as orthorhombic BCF8515. Figure 3 shows the free energy change of reactions (I)–(III) with respect to temperature for the cubic BCF8515 surface containing no oxygen vacancies and for the surface of cubic BCF8515 containing one oxygen vacancy. All the equations of the lines plotted in Figure 3 can be found in S6 of the SI.

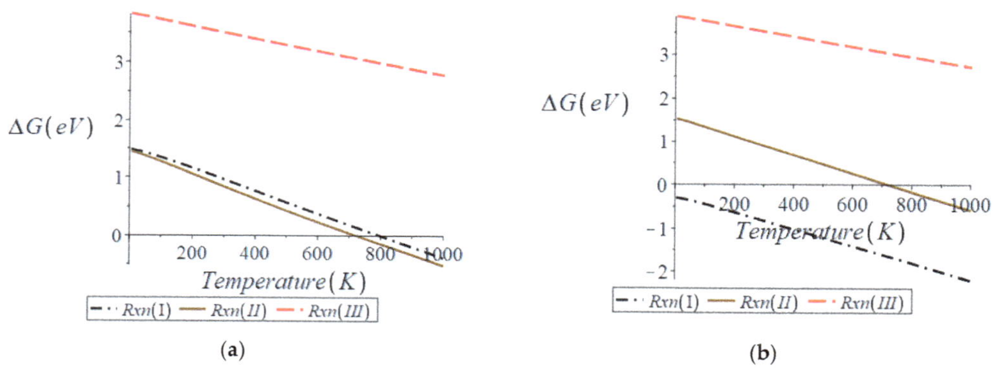

Figure 3. Surface of cubic BCF8515 for (**a**) surface with no oxygen vacancy defects, (**b**) surface with 1 oxygen vacancy defect. Plots were generated in Maple™.

From Figure 3a it can be seen that for the surface containing no oxygen vacancies ("reactant" slab stoichiometry $Ba_{16}Ce_{14}Fe_2H_2O_{48}$), reaction (II) is thermodynamically favored over reactions (I) and (III) over all relevant temperatures. This indicates that the preferred product from the surface containing no oxygen vacancies is H_2O. Since the preferred product from a surface with no oxygen vacancy defects generates an oxygen vacancy defect, the free energy change for reactions (I)–(III) were investigated for a surface containing a single oxygen vacancy ("reactant" slab stoichiometry $Ba_{16}Ce_{14}Fe_2H_2O_{47}$). From Figure 3b it can be seen that the formation of H_2 is thermodynamically preferred over reactions (II) and (III) and is spontaneous over all temperatures. This result indicates that the surface of cubic BCF8515 with an oxygen vacancy defect concentration of no greater than 0.0130 vacancies per $Å^2$ begins to stabilize to reduction in the presence of hydrogen.

3.1.3. Cubic BCF1585

Cheng et al. [10] reported that only BCF8515 transports atomic hydrogen through the binary phase material BCF8515-BCF1585. Cheng et al. [10] and Xia et al. [12] also reported that the surface of BCF8515-BCF1585 was composed of a homogeneous mixture of BCF8515 and BCF1585 grains. The stability of cubic BCF1585 was investigated in the same manner as BCF8515. The results obtained can be found in Figure 4. All the equations of the lines plotted in Figure 4 can be found in S7 of the SI.

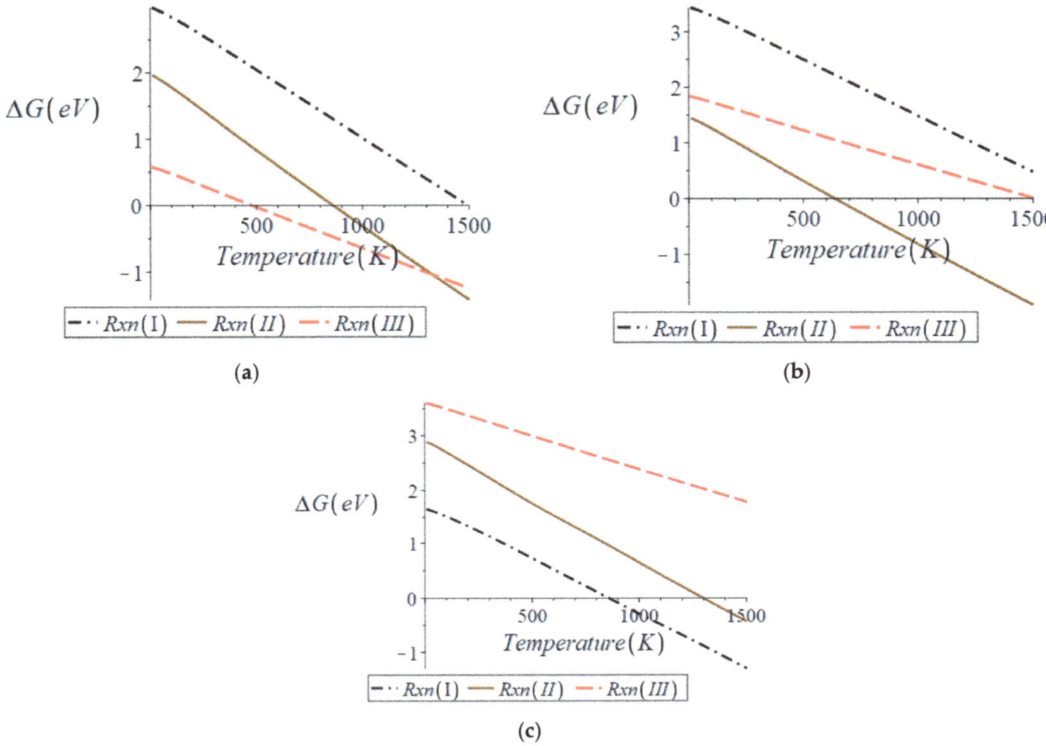

Figure 4. Surface of cubic BCF1585 for (**a**) surface with no oxygen defects, (**b**) surface with 1 oxygen vacancy, (**c**) surface with 2 oxygen vacancies. Plots were generated in Maple^{TM}.

From Figure 4a, it can be seen that the formation of an oxygen vacancy from the oxygen-defect-free surface ("reactant" slab stoichiometry $Ba_{16}Fe_{14}Ce_2H_2O_{48}$) via the loss of $\frac{1}{2}$ O_2 is thermodynamically preferred over the other possible reductions, i.e., reactions (I) and (II). At elevated temperatures, reduction via reaction (II) becomes the thermodynamically favored reaction; however, this does not occur until approximately 1350 K, just above the standard operating temperatures of membrane reactors of 1100 K [26,27]. Since the thermodynamically preferred reaction from the oxygen-defect-free surface generates an oxygen vacancy, reactions (I)–(III) were analyzed from a surface containing one oxygen vacancy ("reactant" slab stoichiometry $Ba_{16}Fe_{14}Ce_2H_2O_{47}$). Figure 4b shows the thermodynamically preferred reaction over all temperatures to be reaction (II). From Figure 4c, it can be seen that once a second oxygen vacancy exists on the surface of BCF1585 ("reactant" slab stoichiometry $Ba_{16}Fe_{14}Ce_2H_2O_{46}$), the formation of H_2 becomes thermodynamically preferred, indicating that the surface is stable to reduction once a surface concentration of at least 0.0222 oxygen vacancies per Å^2 exist on the surface.

The results presented here for BCF1585 indicate that once sufficient oxygen vacancies exist on the surface (a surface vacancy concentration of 0.0222 oxygen vacancies per Å^2), the loss of H_2 becomes the dominant process. The desorption of H_2 from the $Ba_{16}Fe_{14}Ce_2H_2O_{46}$ surface of BCF1585 becomes spontaneous ($\Delta G < 0$) above 859 K. Since it was found that from stabilized surfaces of either BCF8515 the structure desorption of H_2 is spontaneous over all temperatures, H_2 is expected to more readily desorb from BCF8515 crystallites than from BCF1585. In Figure 5 can be found the oxygen defect free structure and structure with the number of oxygen defects where stabilization was found to begin to occur, for each material investigated.

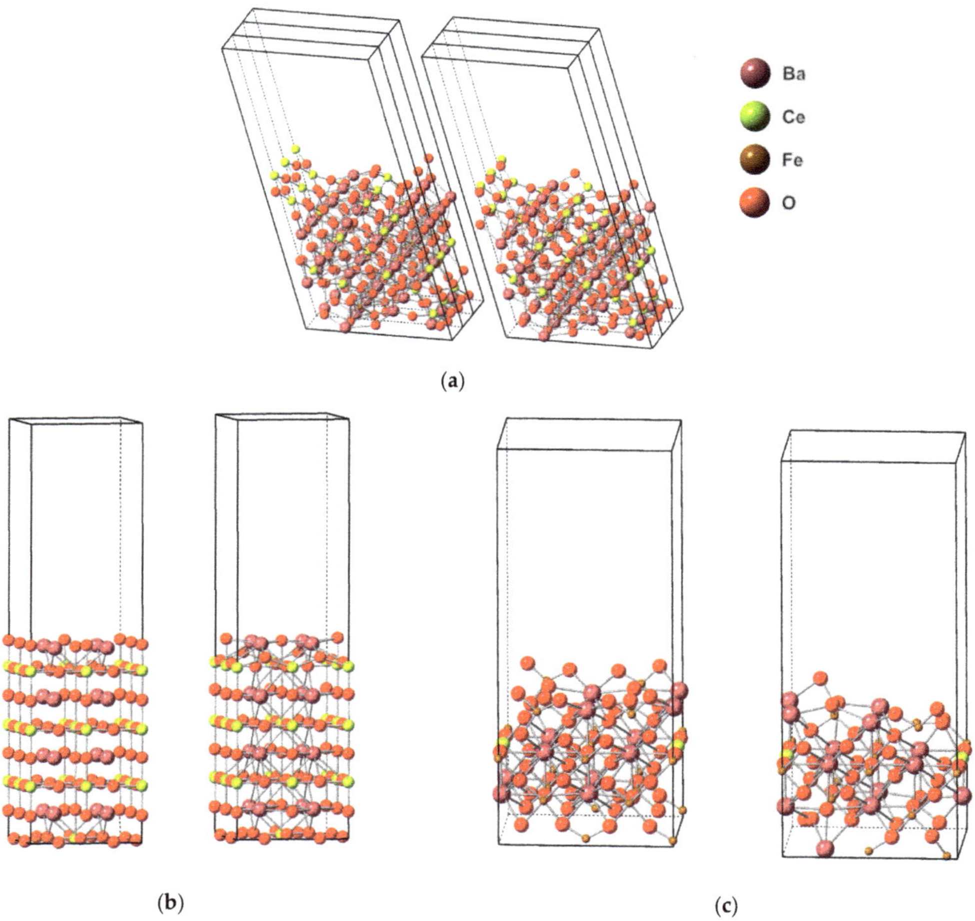

Figure 5. (**a**) Right: orthorhombic BCF8515 with 3 oxygen vacancies on the surface, 3-unit cells in the x-direction. Left: Oxygen-vacancy-defect-free orthorhombic BCF8515 surface, 3-unit cells in the x-direction. (**b**) Right: Cubic BCF8515 with 2 oxygen vacancies on the surface. Left: Oxygen-vacancy-defect-free cubic BCF8515. (**c**) Right: BCF1585 with 3 oxygen vacancies on the surface. Left: Oxygen-vacancy-defect-free BCF1585. Images generated using CrystalMaker®, CrystalMaker Software Ltd., (Oxford, UK), (www.crystalmaker.com, accessed on 26 December 2022).

3.2. *Electron Transfers in the Bulk Due to Reductions*

Reactions of a metal oxide with diatomic hydrogen are colloquially referred to as reductions, because often the transformation results in a decrease in the formal charge of the metal in the material upon decomposition via dehydration [8,9]. The observed stability of the materials examined in this work implies that the reactions of the surface with hydrogen may not simply be reductions. Bonding in solids can be more complicated than the general definitions of covalent and ionic bonds [28], so the chemical transformations of the materials investigated will herein be discussed in terms of changes in coordination number of the metal sites in the material.

The three possible gas-phase products of reactions (I)–(III) are: $H_{2(g)}$, $H_2O_{(g)}$, and $\frac{1}{2}$ $O_{2(g)}$, respectively. H_2 is formed from H atoms on two surface hydroxyl groups. Upon formation of H_2, the coordination number of each of the associated surface O atoms is reduced by one, provided there is no change in the metal–oxygen coordination for any of the nearest neighbor metal atoms. The loss of $\frac{1}{2}$ O_2 is equivalent to ejection of a single oxygen atom. The coordination number of each metal atom that the ejected oxygen was coordinated to is reduced by one, provided the metal coordination number does not further change due to surface reconstruction. When a water molecule is lost from the surface, as with the ejection of a single oxygen atom, the metal coordination number of each metal atom that was coordinated to the lost oxygen will decrease in the absence of surface reconstruction.

In the following discussion, the metal coordination changes are tracked for each transformation resulting from reactions (I)–(III) for each of the three structures considered. The coordination changes are used to qualitatively analyze the structural stability of the surface as a whole. This is accomplished by comparing the relative thermodynamic stability of the surface resultant from reactions (I)–(III) with the change in metal coordination on each surface. The thermodynamic ranking for each of reactions (I)–(III) is well represented by the enthalpy change for each respective reaction. Surface reconstruction was observed for some of the resultant surfaces, so the net change in metal coordination was compared to the thermodynamic stability for each surface. The relevant results can be found in Table 1.

Table 1. Enthalpy change and net metal coordination change for each reaction. Bold values indicate the dominant reaction.

		No Oxygen Vacancies Reactant		**One Oxygen Vacancy Reactant**		**Two Oxygen Vacancy Reactant**	
Material	**Product**	**Net ΔM-O**	**ΔH (Kcal/mol)**	**Net ΔM-O**	**ΔH (Kcal/mol)**	**Net ΔM-O**	**ΔH (Kcal/mol)**
Orthorhombic BCF8515	H_2	+1	82	0	31	**+2**	**13**
	H_2O	**−1**	**60**	**+1**	**26**	−3	36
	$\frac{1}{2}$ O_2	−1	89	−1	72	−5	128
Cubic BCF8515	H_2	0	47	**+1**	**6**	N/A	N/A
	H_2O	**−1**	**34**	−2	37	N/A	N/A
	$\frac{1}{2}$ O_2	+1	88	−3	91	N/A	N/A
BCF1585	H_2	0	84	0	94	**+2**	**53**
	H_2O	−3	50	**−2**	**38**	−2	71
	$\frac{1}{2}$ O_2	**−3**	**16**	−7	45	−1	85

3.2.1. Structural Changes of Orthorhombic BCF8515

Analyzing the structural changes resulting from reactions (I)–(III) on the protonated, defect-free surface of the orthorhombic phase of BCF8515 ($Ba_{16}Ce_{14}Fe_2H_2O_{48}$), it was found that reaction (I) results in a net increase in metal coordination of one, reaction (II) results in a net decrease of one metal coordination, and reaction (III) results in a net metal coordination decrease of one due to the loss of an oxygen. Reaction (II) was identified as the transformation with the smallest enthalpy change. Reaction (I) was determined to have the second largest enthalpy change and reaction (III) was found to have the largest enthalpy change. The most negative free energy change was found to be for the reaction that yielded the smallest enthalpy change, as reactions (I)–(III) were endothermic from all surfaces. Comparison of the structural changes caused by each respective reaction to the stability of the resultant structure indicates a decrease in the metal coordination and the loss of a water from the surface yielded the most thermodynamically favorable reaction. The most likely resultant surface starting from a protonated, oxygen-defect-free surface, was found to contain one oxygen vacancy, no surface hydroxyl groups, and a decrease of one metal coordination.

For reactions (I)–(III) from the orthorhombic BCF8515 surface containing one oxygen vacancy ($Ba_{16}Ce_{14}Fe_2H_2O_{47}$), reaction (I) results in no net change of surface metal coordination, reaction (II) results in an increase of one metal coordination, and reaction (III) results in a net decrease of one metal coordination. Reaction (II) was found to have the smallest enthalpy change. Reaction (I) was found to have the second largest enthalpy change. Reaction (III) was found to have the largest enthalpy change. The most thermodynamically favorable resultant surface due to reactions (I)–(III) from orthorhombic BCF8515 with stoichiometry $Ba_{16}Ce_{14}Fe_2H_2O_{47}$ was found to be the surface with stoichiometry $Ba_{16}Ce_{14}Fe_2O_{46}$, the surface with two oxygen vacancies and no hydroxyl groups, and an increase in metal coordination of one.

Starting from a protonated surface containing two oxygen vacancy defects ($Ba_{16}Ce_{14}Fe_2H_2O_{46}$), reactions (I)–(III) yield a net increase in metal coordination of two for reaction (I), a net decrease in metal coordination of three for reaction (II), and a net decrease of five metal coordinations for reaction (III). Reaction (I) was determined to have the smallest enthalpy change. Reaction (II) was found to have the second largest enthalpy change. Reaction (III) was found to have the largest enthalpy change. Thus, the thermodynamically preferred resultant surface is the result of deprotonation of the two-oxygen-vacancy-containing surface by the formation of H_2, which results in an increase in the metal coordination. In Figure 6 the transformations that result in the thermodynamically preferred resultant surface can be found.

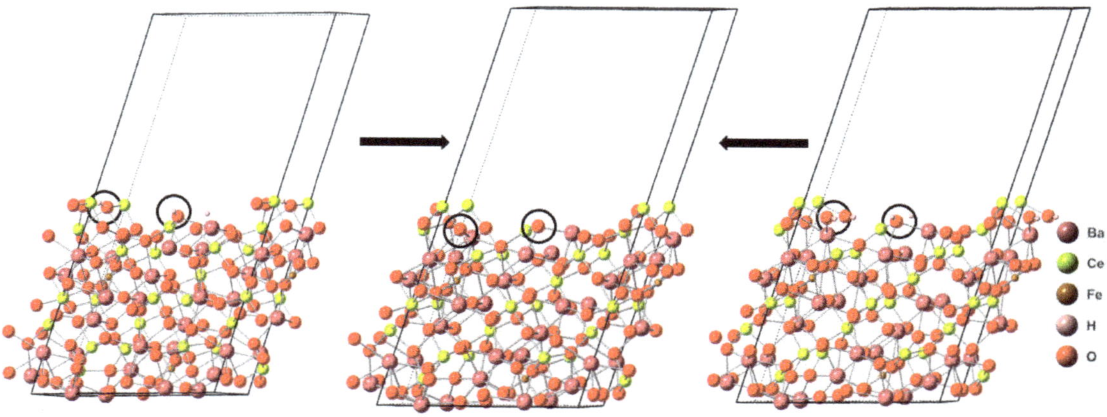

Figure 6. The reaction sites are indicated by black circles. The lefthand structure stoichiometry is $Ba_{16}Ce_{14}Fe_2H_2O_{47}$. The righthand structure stoichiometry is $Ba_{16}Ce_{14}Fe_2H_2O_{46}$. The center structure stoichiometry is $Ba_{16}Ce_{14}Fe_2O_{46}$. Images generated using CrystalMaker®, CrystalMaker Software Ltd., (Oxford, UK), (www.crystalmaker.com, accessed on 26 December 2022).

3.2.2. Structural Changes of Cubic BCF8515

Analyzing structural changes associated with reactions (I)–(III) for the surface of cubic BCF8515 with no surface oxygen vacancies ($Ba_{16}Ce_{14}Fe_2H_2O_{48}$), it was found that reaction (I) results in no net change in metal coordination, reaction (II) results in a net decrease in metal coordination of one, and reaction (III) results in a net increase of one metal coordination. Reaction (II) was found to have the lowest enthalpy change. Reaction (I) had the second lowest enthalpy change and reaction (III) had the highest enthalpy change. The most thermodynamically favorable reaction resulted in the generation of an oxygen vacancy and a net decrease in the metal coordination of the surface.

Starting from the protonated surface containing one oxygen vacancy ($Ba_{16}Ce_{14}Fe_2H_2O_{47}$), reaction (I) results in a net increase of one metal coordination, reaction (II) results in a net decrease in metal coordination of two, and reaction (III) results in a net decrease in metal coordination of three. Reaction (I) was found to have the smallest enthalpy change. Reaction (II) was found to have the second highest enthalpy change, and reaction (III) was found to result in the largest enthalpy change. The most thermodynamically favorable reaction from the protonated surface with one oxygen vacancy resulted in a net increase in metal coordination and the deprotonation of the surface. In Figure 7 the transformations that result in the thermodynamically preferred resultant surface can be found.

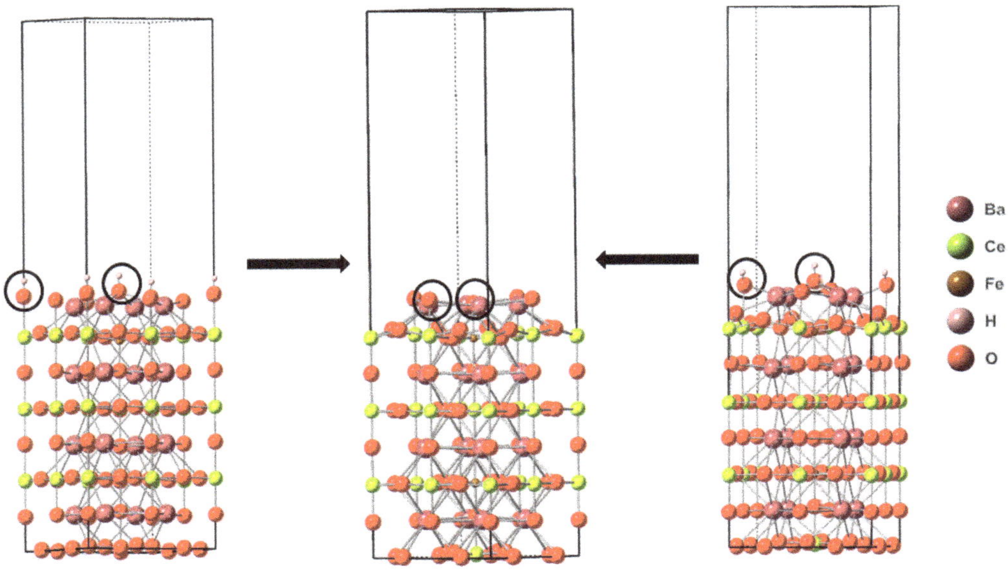

Figure 7. The reaction sites are indicated by black circles. The lefthand structure stoichiometry is $Ba_{16}Ce_{14}Fe_2H_2O_{48}$. The righthand structure stoichiometry is $Ba_{16}Ce_{14}Fe_2H_2O_{47}$. The center structure stoichiometry is $Ba_{16}Ce_{14}Fe_2O_{47}$. Images generated using CrystalMaker®, CrystalMaker Software Ltd., (Oxford, UK), (www.crystalmaker.com, accessed on 26 December 2022).

3.2.3. Structural Changes of BCF1585

Analyzing the structural changes that result from reactions (I)–(III) from the protonated defect-free surface of BCF1585 ($Ba_{16}Fe_{14}Ce_2H_2O_{48}$), it was found that reaction (I) results in no net change in the number of metal coordination, reaction (II) results in a net decrease in metal coordination of three, and reaction (III) results in a net decrease in metal coordination of three. Reaction (III) was found to have the smallest enthalpy change. Reaction (II) was found to have the second smallest enthalpy change and reaction (I) was found to have the largest enthalpy change. The thermodynamically most favorable reaction results in the ejection of an oxygen atom, generating an oxygen vacancy on a still-protonated surface, and a net decrease in the metal coordination.

Starting from a protonated surface with one oxygen vacancy ($Ba_{16}Fe_{14}Ce_2H_2O_{47}$), the following structural changes were determined: Reaction (I) results in no changes in the metal coordination. Reaction (II) results in a net decrease in metal coordination of three. Reaction (III) results in a net decrease in metal coordination of seven. The smallest enthalpy change was determined for reaction (II). The second smallest enthalpy change was determined for reaction (III), and the largest enthalpy change was determined for reaction (I). The thermodynamically most favorable reaction generates a deprotonated resultant surface with an additional oxygen vacancy and a net decrease in metal coordination.

Starting from a protonated surface with two oxygen vacancies ($Ba_{16}Fe_{14}Ce_2H_2O_{46}$), the structural changes resulting from reactions (I)–(III) are as follows: Reaction (I) yields a net increase in metal coordination of two. Reaction (II) results in a net decrease in metal coordination of two. Reaction (III) results in a net decrease in metal coordination of one. Reaction (I) was found to have the smallest enthalpy change, while reaction (II) had the second largest, and reaction (III) had the largest enthalpy change. The most thermodynamically favorable reaction results in a structure that is deprotonated, with no additional creation of oxygen vacancies and results in a net increase in metal coordination. In Figure 8 the transformations that result in the thermodynamically preferred resultant surface can be found.

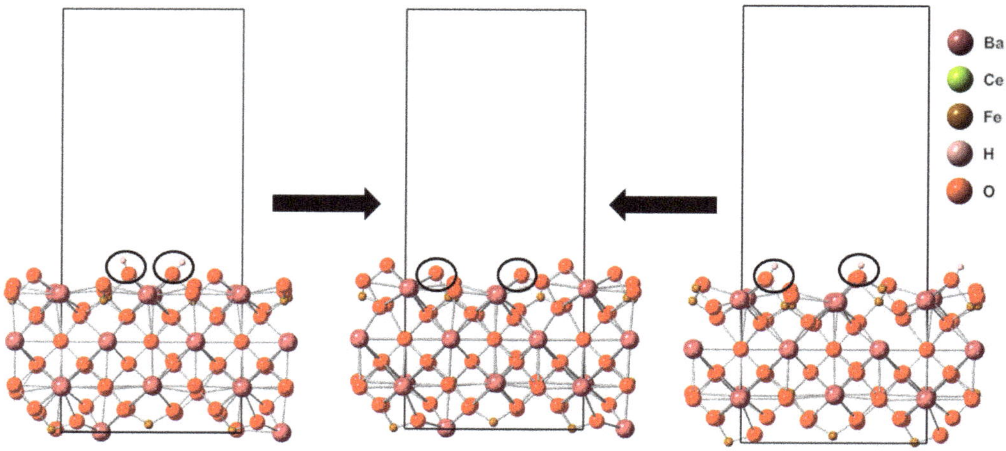

Figure 8. The reaction sites are indicated by white circles. Lefthand structure stoichiometry is $Ba_{16}Ce_{14}Fe_2H_2O_{47}$. The righthand structure stoichiometry is $Ba_{16}Ce_2Fe_{14}H_2O_{46}$. The center structure stoichiometry is $Ba_{16}Ce_2Fe_{14}O_{46}$. Images generated using CrystalMaker®, CrystalMaker Software Ltd., (Oxford, UK), (www.crystalmaker.com, accessed on 26 December 2022).

4. Discussion

All three structures investigated were found to begin to stabilize once a sufficient concentration of oxygen vacancy defects existed on the surface. Both cubic and orthorhombic BCF8515 structures were determined to lose oxygens from pristine (oxygen-defect-free) surfaces via dehydration. The formation of oxygen vacancies via the loss of $\frac{1}{2}$ O_2 from the pristine surfaces of BCF8515 was found to not be spontaneous within relevant temperatures. Cubic BCF8515 was found to begin to stabilize at a stoichiometry no richer in oxygen vacancies than $Ba_{16}Ce_{14}Fe_2H_2O_{47}$. Orthorhombic BCF8515 was found to begin to stabilize at a stoichiometry no richer in oxygen vacancies than $Ba_{16}Ce_{14}Fe_2H_2O_{46}$. BCF1585 was found to preferentially form oxygen vacancies via the loss of $\frac{1}{2}$ O_2 from a pristine surface. From a surface containing one oxygen vacancy per surface cell, the dominant reduction process was found to be the loss of a water. From a surface cell containing two oxygen vacancies, the dominant reduction process for BCF1585 was found to be the loss of H_2. Thus, BCF1585 was found to begin to stabilize at a stoichiometry no richer in oxygen vacancies than $Ba_{16}Fe_{14}Ce_2H_2O_{46}$.

It is well known that perovskite oxides often form with intrinsic oxygen vacancy defects [6,7,10,13,29–31], including BCF8515–BCF1585. For this reason, the stoichiometry for this material is written as $BaCe_xFe_{1\text{-}x}O_{3\text{-}\delta}$, where x is between 0 and 1, and δ is between 0 and 0.5. From the slab stoichiometries for the stable structures presented above, the minimum δ for each structure was found to be 0.063 for cubic BCF8515, and 0.125 for both orthorhombic BCF8515 and BCF1585. The values for δ calculated here represent an upper limit on the minimum number of oxygen vacancies required for the surface to begin to stabilize. From all three surfaces examined, further loss of water is a spontaneous process at elevated temperatures, but the loss of H_2 was found to become the dominant process for the δ values presented. The value of δ determined here for BCF1585 is nearly half the value of δ = 0.22 reported by Zhu et al. [13] for the as-synthesized BCF1585, indicating the material formed with a concentration of oxygen vacancies above the minimum required for stabilization to begin.

In all three materials investigated, it was found that while reactions (II) or (III) are the initially dominant processes (reactions that generate oxygen vacancies) the structural changes all result in a decrease in metal–oxygen bonding, and thus a decrease in the metal coordination in the material. Without surface reconstruction, processes that result in the loss of an oxygen reduce the metal coordination in the material. Once the loss of H_2 becomes the thermodynamically preferred process, an increase in metal coordination (an increase in metal–oxygen bonds) was found to occur in all the materials investigated. These results demonstrate that for a material where the number of oxygen defects is less than the minimum required concentration for stabilization, the metals in the material are electron-poor. Stabilization begins to occur once the metals in the material are in a more favorable (reducing) electronic environment. At the degree of oxygen unsaturation where H_2 desorbing results in an increase in metal coordination, the enthalpy difference between the protonation and deprotonation becomes small enough that H_2 desorption becomes dominant. At an oxygen vacancy concentration where protonation decreases metal coordination, deprotonation increases metal coordination, and loss of an oxygen decreases metal coordination, which is more energetically costly than deprotonation. Once this occurs, deprotonation becomes the dominant process, and the material stabilizes.

A common explanation for the intrinsic oxygen-defect nature of this class of materials is the preferred oxidation state of the B-site metals being 3+ [6,13] (oxygen vacancy defects reduce the Ce or Fe to 3+ from 4+). In $BaCe_xFe_{1-x}O_{3-\delta}$, reducing a significant amount of the Fe or Ce to 3+ would result in a significant loss of oxygen from the material, thus the material would not stabilize as rapidly as determined here. This was observed by Zhu et al. [13] in BCF1585. Clearly, the results presented here indicate that the cause of the stability is not entirely due to the complete reduction of all the B-site metals, or a significantly higher oxygen vacancy concentration would be required before the energy required to remove an oxygen becomes greater than that required for deprotonation. These results encourage further investigations into the specific electronic environment that leads to the perovskite oxide stability observed in reducing atmospheres.

From the results obtained for both BCF1585 and BCF8515, water should be observed desorbing from the permeate surface of the binary material as oxygen vacancies form, provided the material as synthesized has a pristine surface with no oxygen defects (an unlikely real-world scenario). Considering the temperatures utilized in the synthesis methods (for calcination and sintering), and the observation that the material contained oxygen vacancy defects prior to testing in the membrane reactor setup [10,12], it is probable that the dehydration of BCF8515-BCF1585 occurs during synthesis, yielding an already-stabilized material prior to use in a membrane reactor system, providing a plausible explanation for why no water was detected by Cheng et al. [10] or Xia et al. [12] in the permeate sweep gas.

5. Conclusions

This theoretical investigation of the stability of the mixed-phase defect perovskite, BCF8515-BCF1585 in the presence of hydrogen reveals that this materials stability is highly dependent on surface structure and the existence of oxygen vacancies on the surface. For each surface investigated, it was found that upon the formation of surface oxygen vacancies, the surfaces begin to stabilize to reduction in a hydrogen-rich environment. The formation of gas-phase species from the protonated surfaces investigated were all entropy-driven processes, while thermodynamic stability was determined largely by the enthalpic changes for each respective reaction. The stability observed arises from the increasing energy required for the removal of successive oxygens. The energy required to reorganize the surface to mitigate the decrease in metal coordination due to the removal of surface oxygens becomes greater than the entropic benefit gained from removing oxygen or water from the surface. Once surface metal coordination decreases enough to make reconstruction too energetically costly, the loss of diatomic hydrogen from the surface becomes the dominant process. This is further supported by the observation that loss of H_2 from the surface leads to an increase in metal coordination for all the structures investigated. This trend is observed for both the cerium- and iron-dominant phases of the material.

Supplementary Materials: The following supporting information can be downloaded at: https://www.mdpi.com/article/10.3390/molecules28031429/s1, S1: Bulk Structures and k-points; S2: Orthorhombic BCF8515 slab structures and k-points; S3: Cubic BCF8515 slab structures and k-points, S4: BCF1585 slab structures and k-points, S5: Free energy equations for orthorhombic BCF8515, S6: Free energy equations for cubic BCF8515, S7: Free energy equations for BCF1585.

Author Contributions: Conceptualization, B.R. and K.S.; methodology, B.R.; software, B.R.; formal analysis, B.R.; investigation, B.R.; resources, K.S.; data curation, B.R.; writing—original draft preparation, B.R.; writing—review and editing, K.S.; visualization, B.R.; supervision, K.S.; project administration, K.S.; funding acquisition, K.S. All authors have read and agreed to the published version of the manuscript.

Funding: This work was supported in part by ACS-PRF #58323-ND10 for which the authors thank the Donors of the American Chemical Society Petroleum Research Fund.

Institutional Review Board Statement: Not applicable.

Informed Consent Statement: Not applicable.

Data Availability Statement: The data presented in this study are available in this article or in the supporting information.

Acknowledgments: The authors thank Viorel Chihaia for the invitation to participate in this Special Issue.

Conflicts of Interest: The authors declare no conflict of interest.

Sample Availability: Not applicable.

References

1. Nnabuife, S.G.; Ugbeh-Johnson, J.; Okeke, N.E.; Ogbonnaya, C. Present and projected developments in hydrogen production: A technological review. *Carbon Capture Sci. Technol.* **2022**, *3*, 100042. [CrossRef]
2. United States Department of Energy. A National Vision of America's Transition to A Hydrogen Economy—To 2030 and Beyond. Available online: https://www.hydrogen.energy.gov/pdfs/vision_doc.pdf (accessed on 20 January 2022).
3. Ji, M.; Wang, J. Review and comparison of various hydrogen production methods based on costs and life cycle impact assessment indicators. *Int. J. Hydrogen Energy* **2021**, *46*, 38612–38635. [CrossRef]
4. Psarras, P.; Anderson, R.; Gómez-Gualdrón, D.A.; Wilcox, J. Material consequences of hydrogen dissolution in palladium alloys observed from first principles. *J. Phys. Chem. C* **2019**, *123*, 22158–22171. [CrossRef]
5. Gielens, F.C.; Tong, H.D.; Vorstman, M.A.G.; Keurentjes, J.T.F. Measurement and modeling of hydrogen transport through high-flux Pd membranes. *J. Membr. Sci.* **2007**, *289*, 15–25. [CrossRef]
6. Muñoz-García, A.B.; Massaro, A.; Schiavo, E.; Pavone, M. Chapter 1—Tuning perovskite–based oxides for effective electrodes in solid oxide electrochemical cells. In *Solid Oxide-Based Electrochemical Devices*; Lo Faro, M., Ed.; Elsevier Science and Technology: San Diego, CA, USA, 2020; pp. 1–25.
7. Hibino, T.; Mizutani, K.; Yajima, T.; Iwahara, H. Evaluation of proton conductivity in SrCeO3, BaCeO3, CaZrO3 and SrZrO3 by temperature programmed desorption method. *Solid State Ion.* **1992**, *57*, 303–306. [CrossRef]
8. Richardson, J.T.; Scates, R.; Twigg, M.V. X-ray diffraction study of nickel oxide reduction by hydrogen. *Appl. Catal. A Gen.* **2003**, *246*, 137–150. [CrossRef]
9. Sastri, M.V.C.; Viswanath, R.P.; Viswanathan, B. Studies on the reduction of iron oxide with hydrogen. *Int. J. Hydrogen Energy* **1982**, *7*, 951–955. [CrossRef]
10. Cheng, S.; Wang, Y.; Zhuang, L.; Xue, J.; Wei, Y.; Feldhoff, A.; Caro, J.; Wang, H. A dual-phase ceramic membrane with extremely high H2 permeation flux prepared by autoseparation of a ceramic precursor. *Angew. Chem. Int. Ed.* **2016**, *55*, 10895–10898. [CrossRef]
11. Assirey, E.A.R. Perovskite synthesis, properties and their related biochemical and industrial application. *Saudi Pharm. J.* **2019**, *27*, 817–829. [CrossRef] [PubMed]
12. Xia, X.; Zhou, H.; Zhang, Y.; Jiang, H. Innovative steam methane reforming for coproducing CO-free hydrogen and syngas in proton conducting membrane reactor. *AIChE J.* **2019**, *65*, e16740. [CrossRef]
13. Zhu, X.; Cong, Y.; Yang, W. Oxygen permeability and structural stability of BaCe0.15Fe0.85O3$-\delta$ membranes. *J. Membr. Sci.* **2006**, *283*, 38–44. [CrossRef]

14. Go, K.S.; Son, S.R.; Kim, S.D. Reaction kinetics of reduction and oxidation of metal oxides for hydrogen production. *Int. J. Hydrogen Energy* **2008**, *33*, 5986–5995. [CrossRef]
15. Perdew, J.P.; Chevary, J.A.; Vosko, S.H.; Jackson, K.A.; Pederson, M.R.; Singh, D.J.; Fiolhais, C. Atoms, molecules, solids, and surfaces: Applications of the generalized gradient approximation for exchange and correlation. *Phys. Rev. B* **1992**, *46*, 6671–6687. [CrossRef] [PubMed]
16. Kresse, G.; Furthmüller, J. Efficient iterative schemes for ab initio total-energy calculations using a plane-wave basis set. *Phys. Rev. B* **1996**, *54*, 11169–11186. [CrossRef]
17. Kresse, G.; Furthmüller, J. Efficiency of ab-initio total energy calculations for metals and semiconductors using a plane-wave basis set. *Comput. Mater. Sci.* **1996**, *6*, 15–50. [CrossRef]
18. Kresse, G.; Hafner, J. Ab initio molecular dynamics for liquid metals. *Phys. Rev. B* **1993**, *47*, 558–561. [CrossRef]
19. Kresse, G.; Joubert, D. From ultrasoft pseudopotentials to the projector augmented-wave method. *Phys. Rev. B* **1999**, *59*, 1758–1775. [CrossRef]
20. Fischer, C.C.; Tibbetts, K.J.; Morgan, D.; Ceder, G. Predicting crystal structure by merging data mining with quantum mechanics. *Nat. Mater.* **2006**, *5*, 641–646. [CrossRef]
21. Opahle, I.; Madsen, G.K.H.; Drautz, R. High throughput density functional investigations of the stability, electronic structure and thermoelectric properties of binary silicides. *Phys. Chem. Chem. Phys.* **2012**, *14*, 16197–16202. [CrossRef]
22. Knight, K.S. Structural phase transitions in BaCeO3. *Solid State Ion.* **1994**, *74*, 109–117. [CrossRef]
23. Wolverton, C.; Hass, K.C. Phase stability and structure of spinel-based transition aluminas. *Phys. Rev. B* **2000**, *63*, 024102. [CrossRef]
24. McQuarrie, D.A. *Statistical Thermodynamics*; University Science Books: Sausalito, CA, USA, 1973.
25. Cai, S.; Caldararu, M.; Sohlberg, K. Entropic contributions to the atomic-scale charge-carrier/surface interactions that govern macroscopic surface conductance. *J. Phys. Chem. C* **2010**, *114*, 3991–3997. [CrossRef]
26. Gallucci, F.; Comite, A.; Capannelli, G.; Basile, A. Steam reforming of methane in a membrane reactor: An industrial case study. *Ind. Eng. Chem. Res.* **2006**, *45*, 2994–3000. [CrossRef]
27. Hafeez, S.; Al-Salem, S.M.; Manos, G.; Constantinou, A. Fuel production using membrane reactors: A review. *Environ. Chem. Lett.* **2020**, *18*, 1477–1490. [CrossRef]
28. Levitin, V. *Interatomic Bonding in Solids: Fundamentals, Simulation, and Applications*; Wiley-VCH Verlag GmbH & Co. KGaA: Weinheim, Germany, 2013.
29. Tao, Z.; Bi, L.; Zhu, Z.; Liu, W. Novel cobalt-free cathode materials $BaCe_xFe_{1-x}O_{3-\delta}$ for proton-conducting solid oxide fuel cells. *J. Power Sources* **2009**, *194*, 801–804. [CrossRef]
30. Yamanaka, S.; Fujikane, M.; Hamaguchi, T.; Muta, H.; Oyama, T.; Matsuda, T.; Kobayashi, S.-I.; Kurosaki, K. Thermophysical properties of $BaZrO_3$ and $BaCeO_3$. *J. Alloys Compd.* **2003**, *359*, 109–113. [CrossRef]
31. Zhu, X.; Wang, H.; Yang, W. Novel cobalt-free oxygen permeable membrane. *Chem. Commun.* **2004**, *9*, 1130–1131. [CrossRef]

molecules

MDPI

Article

Molecular Dynamics Investigation of Wettability Alteration of Quartz Surface under Thermal Recovery Processes

Mohammadali Ahmadi * and Zhangxin Chen

Department of Chemical and Petroleum Engineering, Schulich School of Engineering, University of Calgary, Calgary, AB T2N 1T4, Canada
* Correspondence: mohammadali.ahmadi@ucalgary.ca

Abstract: One of the primary methods for bitumen and heavy oil recovery is a steam-assisted gravity drainage (SAGD) process. However, the mechanisms related to wettability alteration under the SAGD process still need to be fully understood. In this study, we used MD simulation to evaluate the wettability alteration under a steam injection process for bitumen and heavy oil recovery. Various oil droplets with different asphaltene contents were considered to determine the effect of an asphaltene content on the adsorption of the oil droplets onto quartz surfaces and wettability alteration. Based on the MD simulation outputs, the higher the asphaltene content, the higher the adsorption energy between the bitumen/heavy oil and quartz surfaces due to coulombic interactions. Additionally, the quartz surfaces became more oil-wet at temperatures well beyond the water boiling temperature; however, they were extremely water-wet at ambient conditions. The results of this work provide in-depth information regarding wettability alteration during in situ thermal processes for bitumen and heavy oil recovery. Furthermore, they provide helpful information for optimizing the in situ thermal processes for successful operations.

Keywords: heavy oil; contact angle; porous media; molecular dynamics; thermal recovery

1. Introduction

The world's primary and vital energy resources are still fossil fuels, including conventional and unconventional oil and gas reservoirs [1,2]. Heavy oil and bitumen have a large share of fossil fuels, but due to their high viscosity, their recovery needs more energy to be injected compared to conventional oil [3,4]. In situ thermal recovery methods, especially steam-assisted gravity drainage (SAGD), are used to extract heavy oil and bitumen [5–7]. Although the SAGD process is a typical and mature recovery method for them, its mechanisms related to wettability alteration still need to be fully understood [8].

Modern research uses molecular dynamics (MD) simulations to assess complex interactions from an atomistic perspective, providing in-depth knowledge of static and dynamic features of such phenomena at the atomic level [9–20]. For example, Zhong et al. [21] used MD simulations to study the adsorption of crude oil molecules onto water-wet quartz surfaces. They showed that more polarized components can infiltrate a water layer and subsequently adsorb to the quartz surfaces. On the other hand, nonpolar components cannot behave like polar components. They further suggested a two-step adsorption approach in light of their findings. Their concept states that polar oil contents adhere to a rock surface in the first stage, creating space and acting as an anchor for polar and nonpolar elements to attach.

Additionally, adsorption and desorption processes in porous media can be explained by specific mechanisms using MD simulations [22]. MD simulations were used by Du et al. [23] to describe how a cationic surfactant interacts with talc. They demonstrated how the hydrophobicity of a talc surface might be enhanced by a DTAB surfactant, creating a bridge to an edge of the surface. To assess the interactions between the cationic, anionic, and muscovite surfaces of a surfactant combination, Wang et al. [24] employed MD simulation. According to their findings, the key elements in the binding interactions between muscovite and the surfactant mixture are the hydrophobic contacts between surfactant tails and the electrostatic interactions between polar head groups. To grasp how silica nanoparticles' wettability can improve gas recovery from nanopores, Sepehrnia and Mohammadi [25] used MD simulations. As a result of nanoparticles adhering to a pore wall, their simulated results showed that hydrophobicity increased. It then had an effect of diminishing liquid bridges within nanopores.

To explore the underlying processes influencing the oil recovery from calcite rock during low-salinity water injection, Zhao et al. [26] performed MD simulations. Non-equilibrium MD (NEMD) simulations were also performed to focus on dynamic oil characteristics. They discovered that increasing water's salt content had no effect on wettability when an environment was oil-wet, but when the environment was partially water-wet, it stimulated wettability toward fully water-wet. They revealed that decane molecules had a propensity to cluster and travel through nanopores as a group. Guo et al. [27] used MD models and laboratory investigations to understand how the quantity and concentration of hydrophilic groups in non-anionic surfactants impacted lignite wettability. According to the findings of their simulations and experiments, they concluded that a linear polyoxy-ethylene ether surfactant has weaker hydrophilic groups than a more effective polyhydroxy surfactant, which allows for a reduction in the hydrophilicity of a lignite surface.

To the best of the authors' knowledge, no work has been published in the literature using MD simulations to evaluate the effect of asphaltene content on the wettability alteration during in situ thermal recovery processes. Contact angle calculations and adsorption energy of heavy oil droplets onto quartz surfaces were employed to analyze MD trajectories. The subsequent sections explain the MD workflow and details of different simulation configurations. The output of this work provides fundamental information about the impact of heavy oil composition and asphaltene content on the wettability alteration during in situ thermal recovery methods.

2. Results and Discussion

Figure 1 compares the contact angles calculated for each system at 298 K and 498 K along with some snapshots of heavy oil droplets on the quartz surface. As shown in this figure, at a temperature of 298 K, the contact angles are well beyond 105°, which reveals a water-wet condition of the quartz surface; however, increasing the asphaltene content of the heavy oil, the contact angle reduced. In other words, increasing the asphaltene content resulted in slight wettability alteration toward oil-wet at 298 K. On the other hand, at a higher temperature (498 K), the contact angle became much less than 75°, which shows a higher tendency of the quartz surface toward oil-wet conditions. Additionally, as illustrated in Figure 1, increasing the asphaltene content can intensify the wetness of the quartz surface toward oil-wet. This observation is attributed to higher adsorption energy between heavy oil droplets and the quartz surface. Our MD simulation outputs are in agreement with experimental observations reported by Naser et al. [28]. It is worth highlighting that the contact angles for heavy oil with 8% and 12% asphaltene contents are similar, which uncovers that at a steam chamber temperature of 498 K, increasing asphaltene content to a specific concentration can reduce the contact angle of the oil significantly, but beyond that concentration, there is no meaningful difference in terms of the contact angle with the quartz surface. The size and structure of asphaltene and resin molecules play a significant role in the colloidal and adsorption behavior of asphaltene molecules in an oil droplet.

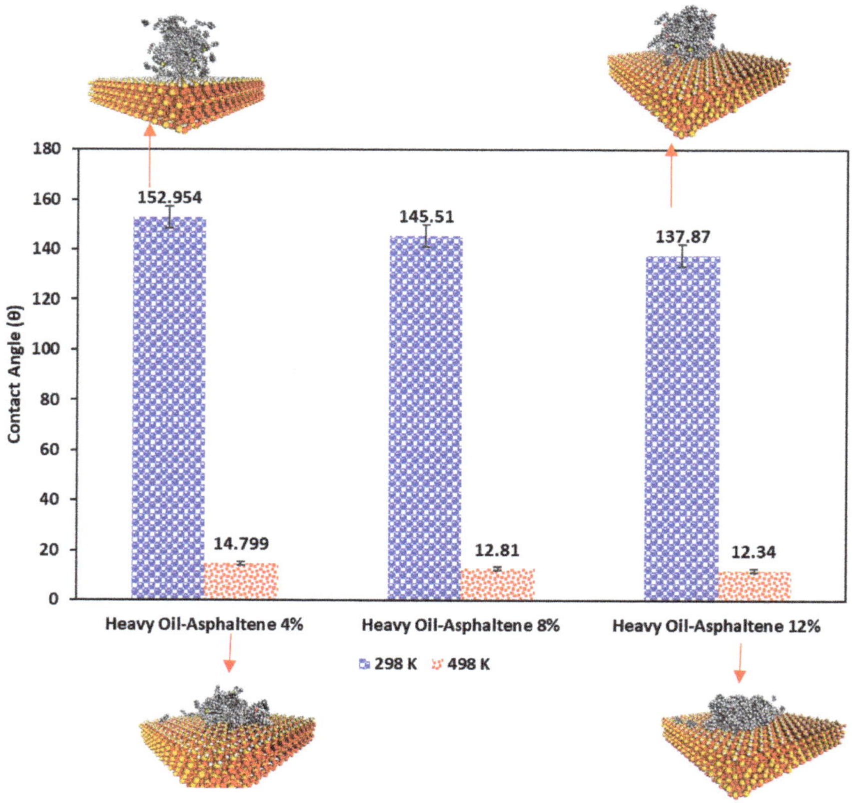

Figure 1. Comparison between the contact angles in each system and different temperatures (for the clarity purposes, water molecules in snapshots are hidden).

The van der Waals interactions are largely determined by the surface area of a molecule and its electronic polarizability. The larger the surface area of a molecule, the more likely it is to have the van der Waals interactions. Electronic polarizability, which describes how easily an electron cloud around an atom can be distorted, is affected by the number of electrons and the size of an atom. Generally, larger molecules have more electrons and are, therefore, more polarizable. The coulombic potential energy is influenced by the polarity of interacting molecules, which can be represented by their dipole moment. The greater the difference in electronegativity between the atoms in a molecule, the larger the dipole moment will be. The presence of heteroatoms can also significantly increase the polarity of a molecule [29–31]. Thus, due to the presence of heteroatoms in the molecular structures of asphaltene and resin, these molecules are more polar than saturates and aromatics. It is also worth noting that the numbers of heptane, toluene, and resin molecules for all the three samples are the same and only the number of asphaltene molecules is different. Hence, the changes in both the van der Waals (VdW) and coulombic interactions between heavy oil droplets and a quartz surface mainly come from asphaltene content variations of the three heavy oil samples. Table 1 reports the adsorption energies between the oil droplets and quartz surface for each scenario at different temperatures. It also reports the VdW and coulombic interaction components of the adsorption energy in each scenario. This table illustrates that the adsorption energy between the heavy oil droplets and quartz surface increased by increasing the system's temperature. Additionally, the coulombic interaction between the heavy oil and quartz is much higher than the van der Waals interaction, which reveals that the polar molecules inside the heavy oil are contributing to the adsorption process of heavy oil droplets onto the quartz surface. In other words, due to the presence of heteroatoms in the molecular structure of asphaltene and resin, these molecules are more polar than saturates and aromatics and interact with the quartz surface via coulombic interactions.

Table 1. Adsorption energy between heavy oil droplets and quartz at 298 K and 498 K.

Temperature (K)	$E_{Adsorption}$ (Kcal/mol)	E_{vdw} (Kcal/mol)	$E_{coulombic}$ (Kcal/mol)
Heavy Oil with 4% Asphaltene			
298	−1308.71	−285.65	−991.33
498	−2430.74	−419.99	−1929.02
Heavy Oil with 8% Asphaltene			
298	−1423.86	−306.18	−1096.85
498	−2580.69	−488.59	−2061.27
Heavy Oil with 12% Asphaltene			
298	−1484.39	−320.82	−1144.11
498	−2762.26	−533.33	−2239.47

Figure 2 demonstrates the percentage of interaction energy difference due to increasing the temperature from 298 K to 498 K. As shown in this figure, increasing the temperature resulted in increasing the adsorption energy by more than 80% due to the extensive contribution of coulombic interactions since asphaltene molecules are polar. Moreover, increasing the asphaltene content can also increase VdW interactions, but the contribution of the VdW interactions is significantly lower than that of the coulombic interactions.

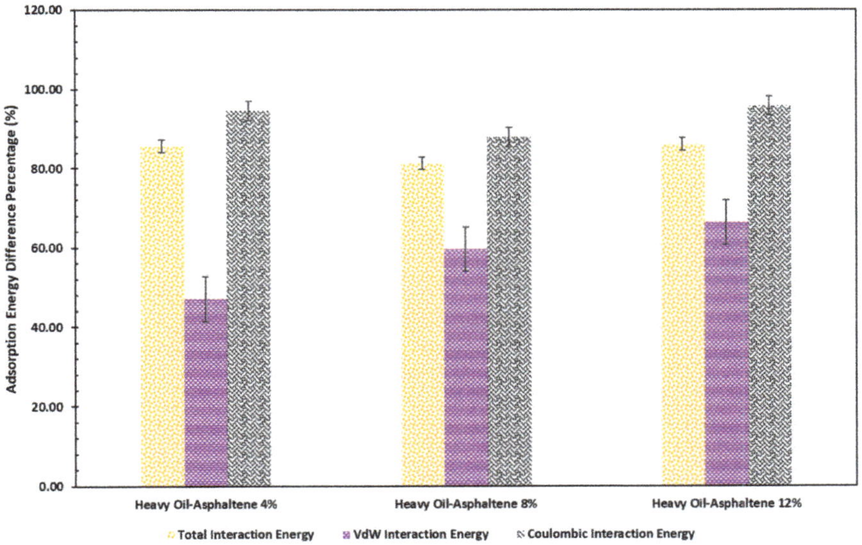

Figure 2. Comparison between the percentage of adsorption energy difference relative to the ambient temperature for each system due to the temperature change from 298 K to 498 K.

The understanding of the mechanism of wettability alteration under the SAGD process can help to optimize the in situ thermal processes for successful operations. The current study has practical and industrial ramifications in heavy oil recovery, as rock wettability can greatly affect the efficiency of these processes. The results of this study can benefit the optimization of the SAGD process and reservoir simulation in several ways. Firstly, the study provides insights into the mechanisms of wettability alteration during the SAGD process, which can be used to optimize the process and improve recovery efficiency. Secondly, the study found that the asphaltene content in oil droplets affects the adsorption energy between the droplets and the quartz surfaces. This information can be used to optimize the SAGD process for different types of heavy oil reservoirs with varying asphaltene contents. Thirdly, this study revealed that the quartz surfaces became more oil-wet at high temperatures, which can be used as input for reservoir simulation to predict an oil recovery rate and optimize the SAGD process. Lastly, the understanding of the wettability alteration during the SAGD process can be used to design a better SAGD process and improve its oil recovery rate in the field. Additionally, the results of this study are specific to the quartz surfaces, and further research may be needed to understand the wettability alteration on different rock types and fluid types. Overall, this work provides valuable information on the effect of thermal recovery processes on the wettability of quartz surfaces, which can be used to optimize the SAGD process and improve heavy oil recovery.

3. Methodology

3.1. Force Field and Simulation Initialization

The Materials Studio software [32] is used to carry out all MD simulations, and the COMPASS III [33–35] force field is applied. Various publications in the literature confirm the COMPASS force field's [33,36–40] ability to represent complicated systems made up of different materials [34,41–45]. For instance, solvation of asphaltene in supercritical water, oil detachment from quartz nanochannels [46–48], and carbon dioxide capturing by nanotube–asphalt composite [49]. The typical average temperature in a steam chamber in thermal heavy oil recovery is about 225 °C or 498 K [50,51]. Thus, to evaluate the impact of temperature on the wettability of a quartz surface, two temperatures, 298 K and 498 K, were employed in MD simulations for each system. Table 2 reports the composition of the heavy oil droplets in terms of the numbers of heptane, toluene, resin, and asphaltene molecules. Oil molecules were randomly placed in a $4 \times 4 \times 4$ nm simulation box to create oil droplets. A total of 7000 water molecules were also added to the created oil droplet to generate an oil droplet in a slab filled by water molecules. Then, geometry optimization was performed on the generated oil and water system. The smart method for 2.0×10^6 steps with the energy cutoff of 2.0×10^{-5} kcal/mol and cutoff displacement of 10^{-5} Å was employed during the geometry optimization process. Then, 2×10^6 steps of an NPT ensemble with one femtosecond (fs) time step were applied to the simulation box filled by the water molecules and oil droplet to achieve a reasonable density. It is worth noting that this process was performed for each temperature (298 K and 498 K), and the pressure was set to 1 MPa. The generated oil droplet and water molecules were placed onto the quartz surface. In the next step, geometry optimization was applied to the entire simulation box to achieve the minimum energy level. It is worth noting that the geometry optimization setting was similar to the previous stage. In the equilibration stage, we conducted 5×10^6 steps of an NVT ensemble with one fs time step to ensure that the simulation box is equilibrated. Sampling was performed after every 5000 steps of simulation for both the equilibration and production stages. Finally, in the MD production stage, we used 5×10^6 steps of the NVT ensemble with a similar time step to analyze the trajectories and calculate the contact angles and adsorption energies. Figure 3 demonstrates the variables, functions, ensembles, and duration of MD simulations, along with their corresponding values. Figure 4 depicts the schematic of creating an MD simulation box to measure a contact angle between a droplet of heavy oil and a water-wet quartz surface. To figure out the precision of the calculated mean values of contact angle and adsorption energy, error bars as the standard errors were employed, which are calculated as the standard deviation of the sample divided by the square root of the sample size.

Table 2. Composition of the heavy oil droplets for each system.

	Heptane	Toluene	Resin	Asphaltene
Heavy Oil ID	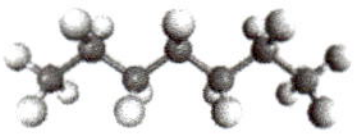	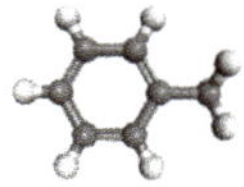	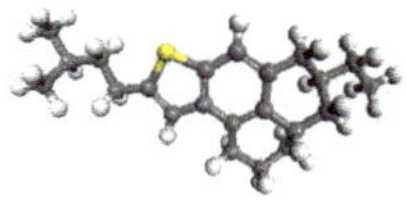	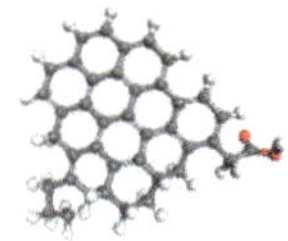
4% Asphaltene	120	120	30	3
8% Asphaltene	120	120	30	5
12% Asphaltene	120	120	30	8

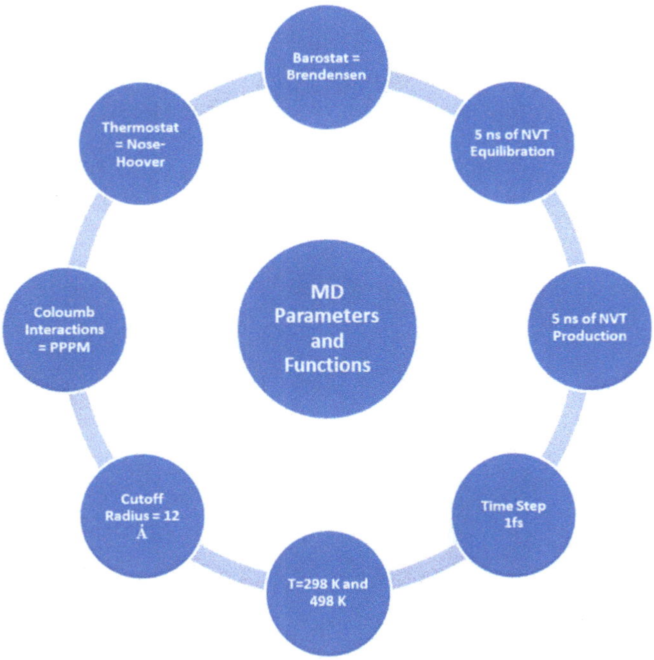

Figure 3. Parameters and MD simulation functions.

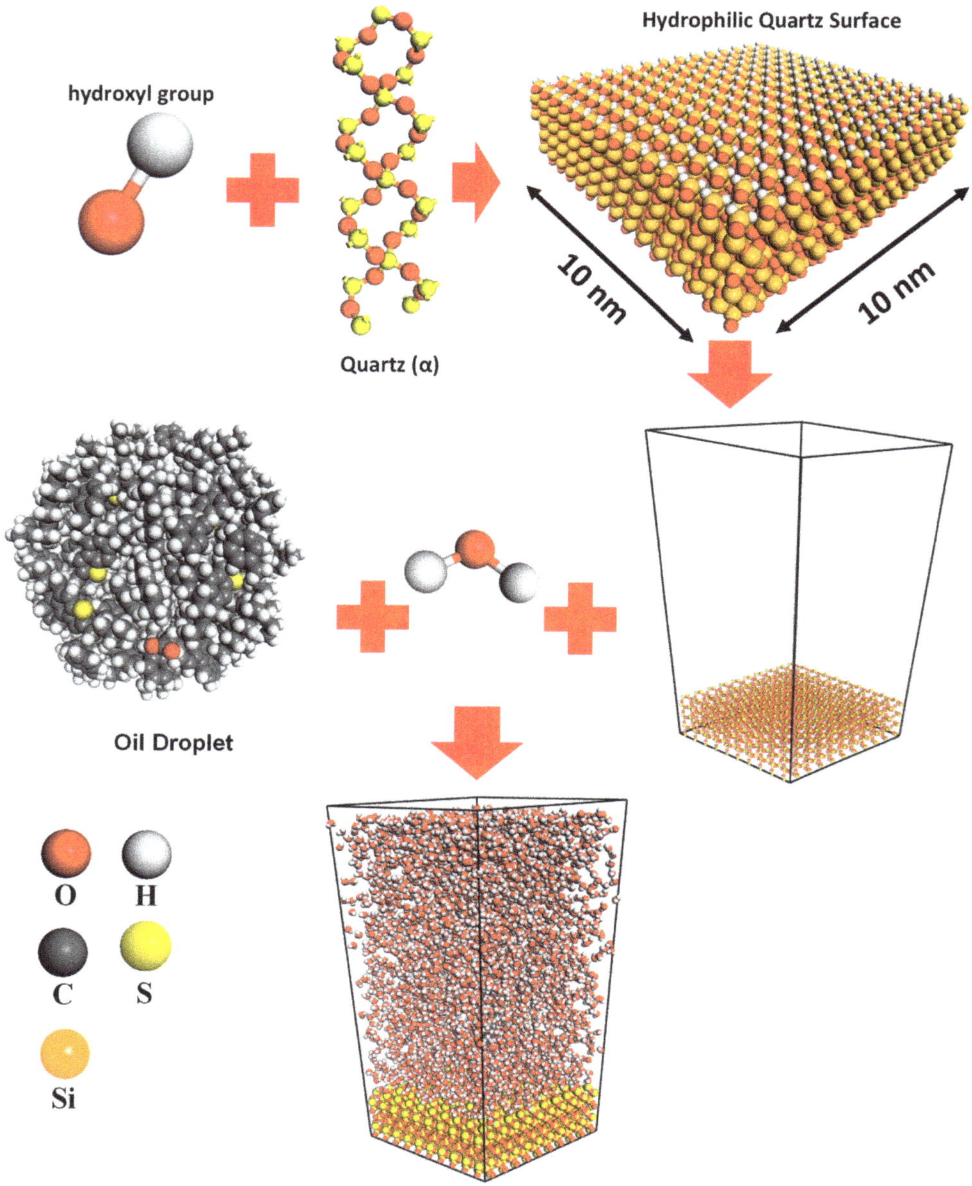

Figure 4. Schematic of simulation box creation to measure a contact angle between a heavy oil droplet on a water-wet quartz.

3.2. Contact Angle Calculations

The wettability of a mineral surface can be quantitatively characterized by a contact angle between a fluid and the mineral surface. Since the parameter of the contact angle cannot be directly obtained in the Materials Studio software, the calculations of the contact angle in this study refer to the Hautman and Klein method [52] through Perl scripting. The Perl Script code is available on request. Moreover, its solution idea is to calculate a microscopic contact angle by comparing the average height of the centroid of an oil droplet

<Z_{com}> with the average height of an ideal spherical droplet intersecting the surface of a calcite. Figure 5 is a schematic diagram of calculating the microscopic contact angle in a molecular simulation. Here, assume that the density of the ideal droplet is homogeneous and equal to the bulk density of oil; that is, the volume of a sphere in the half-space on the surface is equal to the volume of the equivalent water droplet, and the position of the center of mass determines the position of the sphere relative to the plane. Therefore, the angle between the surface of the ideal spherical droplet in Figure 5 and the plane is the contact angle (θ) of the water droplet, and the relationship between it and the average height of the droplet centroid <Z_{com}> can be expressed as follows:

$$\langle Z_{com} \rangle = (2)^{-\frac{4}{3}} R_o \left(\frac{1-cos\theta}{2+cos\theta} \right)^{\frac{1}{3}} \left(\frac{3+cos\theta}{2+cos\theta} \right) \tag{1}$$

where the average height of the droplet centroid <Z_{com}> is the height based on the surface of a slab, and the radius of a droplet can be calculated by the total number of molecules (N) in the water droplet, which can be expressed as $R_o = \left(\frac{3N}{4\pi\rho_o} \right)^{\frac{1}{3}}$. Based on the above method, the microscopic contact angle of water droplets on a calcite surface can be obtained in MS software. It should be noted that a rock surface is predominately water-wet/steam-wet when the contact angle is between 105° and 180°. The rock surface is predominantly oil-wet between 0 and 75 degrees. The contact angle for intermediate-wet rock is between 75° and 105° [28,53].

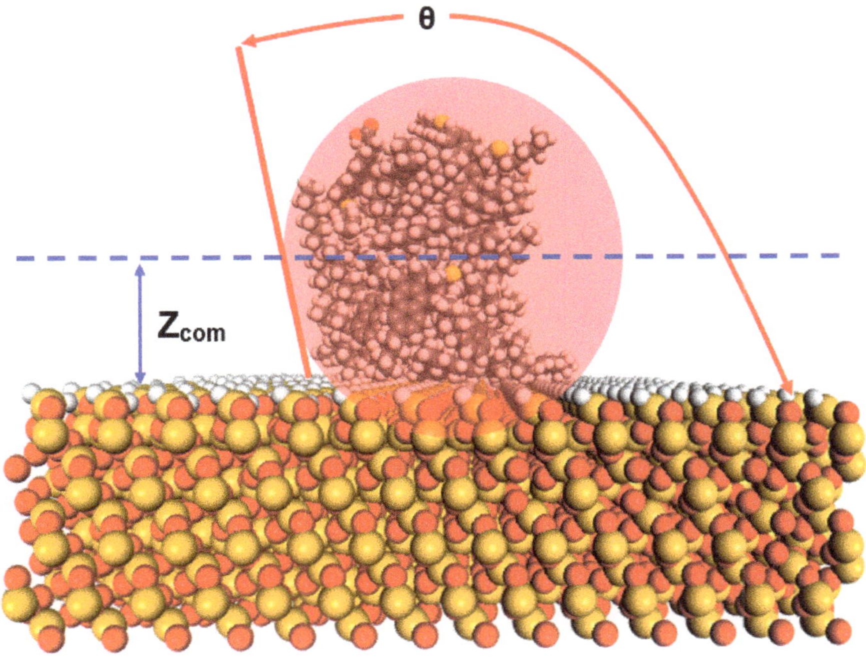

Figure 5. Graphical illustration of the contact angle measurement between heavy oil droplets and a quartz surface.

3.3. Adsorption Energy (AE)

Adsorption energy and interaction energy are closely related but different concepts in MD simulations. Adsorption energy is the energy required for a molecule to adhere to a surface and is a measure of the strength of the interaction between the adsorbate (a droplet of heavy oil) and the adsorbent (a quartz surface). If the energy required is negative, it means the adsorbate is stable on the surface, while a positive energy means the adsorbate is not stable. On the other hand, the total potential energy is the sum of the potential energy of each atom in the system, whereas the interaction energy is the sum of the potential energy of each adsorbate–adsorbent pair. In MD simulations, the process of determining the adsorption energy of a molecule on a surface involves a few key steps. First, the initial configuration of a system must be established such that an adsorbate is in contact with the surface of an adsorbent. The total energy of the system, including both the adsorbate and adsorbent, must then be calculated utilizing a chosen force field. Next, the interactions between the adsorbate and the adsorbent must be quantified by summing up the potential energy of each atom in the system. By subtracting the potential energy of the isolated adsorbate and adsorbent from the total energy, the adsorption energy can be determined. The system must be equilibrated for a sufficient period of time before the adsorption energy is calculated to ensure that the system is in thermodynamic equilibrium. After that, a production run should be performed, and the data collected during this time can be processed to calculate the adsorption energy. It is worth noting that this approach assumes that the system is in thermal equilibrium and that the adsorbate–adsorbent interactions are represented by classical potentials, and it may not be able to capture quantum mechanical effects that can influence the adsorption process. To analyze an adsorption process of heavy oil onto a quartz surface, we used the interaction energy between heavy oil droplets and the quartz surface inside a simulation box. The following equation determines our system's adsorption energy between oil and quartz [54]:

$$E_{Adsorption} = E_{Total} - \left(E_{heavy\ oil} + E_{Silica}\right) \tag{2}$$

where $E_{Adsorption}$ (kcal/mol) stands for the adsorption energy between the heavy oil droplets and quartz, E_{Total} (kcal/mol) represents the total energy of the heavy oil droplets and quartz, $E_{heavy\ oil}$ (kcal/mol) denotes the energy of the heavy oil droplets inside the system, and E_{Silica} (kcal/mol) represents the energy of the quartz inside the system [54].

4. Conclusions

In this work, we used MD simulations to assess how the wettability of a quartz surface can change under SAGD operation conditions. Additionally, to figure out the effect of asphaltene concentration on the adsorption of heavy oil droplets onto the quartz surface and a wettability change, several heavy oil droplets with varying concentrations of asphaltene were considered. The following conclusions can be drawn from the findings of this study's MD simulations:

- Increasing the temperature from the ambient temperature (298 K) to a typical steam chamber temperature (498 K) can significantly increase the adsorption energy between heavy oil droplets and the quartz surface. In other words, heavy oil droplets tend to adsorb onto the quartz surface at higher temperatures than ambient conditions.
- Increasing the asphaltene content of the heavy oil can also increase the adsorption energy significantly, and this behavior was intensified at a higher temperature. However, the rate of change of the adsorption energy is not linearly correlated with the asphaltene content of the heavy oil.
- According to the adsorption energy calculations, a coulombic interaction was the main contributor to an adsorption process due to the polarity of the asphaltene and resin molecules since they have heteroatoms in their structures.
- Based on the contact angle calculations, at ambient temperature, a quartz surface tends to be water-wet due to the lower adsorption energy of the heavy oil at that temperature; however, at temperatures well beyond the water boiling temperature, the quartz surface became more oil-wet thanks to higher energy adsorption.
- The results of this study are specific to the quartz surfaces and one type of asphaltene and resin; further research may be needed to understand the wettability alteration on different rock types and various types of asphaltenes and resins.

Author Contributions: Conceptualization, M.A.; methodology, M.A.; formal analysis, M.A.; investigation, M.A. writing—original draft preparation, M.A.; writing—review and editing, M.A. and Z.C.; supervision, Z.C. All authors have read and agreed to the published version of the manuscript.

Funding: NSERC-Vanier Canada Graduate Scholarships (CGS) and Izaak Walton Killam Memorial Scholarship funded this research. It is also partly supported by NSERC/Energi Simulation Chair.

Data Availability Statement: Data available on request from the authors.

Conflicts of Interest: The authors declare no conflict of interest.

References

1. Mohr, S.; Wang, J.; Ellem, G.; Ward, J.; Giurco, D. Projection of world fossil fuels by country. *Fuel* **2015**, *141*, 120–135. [CrossRef]
2. Ahmadi, M.; Chen, Z. Challenges and future of chemical assisted heavy oil recovery processes. *Adv. Colloid Interface Sci.* **2020**, *275*, 102081. [CrossRef] [PubMed]
3. Shah, A.; Fishwick, R.; Wood, J.; Leeke, G.; Rigby, S.; Greaves, M. A review of novel techniques for heavy oil and bitumen extraction and upgrading. *Energy Environ. Sci.* **2010**, *3*, 700–714. [CrossRef]
4. Huc, A.-Y. *Heavy Crude Oils: From Geology to Upgrading: An Overview*; Technip: Paris, France, 2010.
5. Guo, K.; Li, H.; Yu, Z. In-situ heavy and extra-heavy oil recovery: A review. *Fuel* **2016**, *185*, 886–902. [CrossRef]
6. Dong, X.; Liu, H.; Chen, Z.; Wu, K.; Lu, N.; Zhang, Q. Enhanced oil recovery techniques for heavy oil and oilsands reservoirs after steam injection. *Appl. Energy* **2019**, *239*, 1190–1211. [CrossRef]
7. Sheng, J. *Enhanced Oil Recovery Field Case Studies*; Gulf Professional Publishing: Houston, TX, USA, 2013.
8. Yang, S.; Nie, Z.; Wu, S.; Li, Z.; Wang, B.; Wu, W.; Chen, Z. A critical review of reservoir simulation applications in key thermal recovery processes: Lessons, opportunities, and challenges. *Energy Fuels* **2021**, *35*, 7387–7405.
9. Hu, H.; Li, X.; Fang, Z.; Wei, N.; Li, Q. Small-molecule gas sorption and diffusion in coal: Molecular simulation. *Energy* **2010**, *35*, 2939–2944. [CrossRef]
10. Rao, Z.; Wang, S.; Peng, F.; Zhang, W.; Zhang, Y. Dissipative particle dynamics investigation of microencapsulated thermal energy storage phase change materials. *Energy* **2012**, *44*, 805–812. [CrossRef]
11. Raaen, S.; Ramstad, A. Monte-Carlo simulations of thermal desorption of adsorbed molecules from metal surfaces. *Energy* **2005**, *30*, 821–830. [CrossRef]
12. Kjelstrup, S.; Bedeaux, D.; Inzoli, I.; Simon, J.-M. Criteria for validity of thermodynamic equations from non-equilibrium molecular dynamics simulations. *Energy* **2008**, *33*, 1185–1196. [CrossRef]
13. Chen, L.; Wang, S.; Tao, W. A study on thermodynamic and transport properties of carbon dioxide using molecular dynamics simulation. *Energy* **2019**, *179*, 1094–1102. [CrossRef]
14. Klonos, P.A.; Goncharuk, O.V.; Pakhlov, E.M.; Sternik, D.; Deryło-Marczewska, A.; Kyritsis, A.; Gun'ko, V.M.; Pissis, P. Morphology, molecular dynamics, and interfacial phenomena in systems based on silica modified by grafting polydimethylsiloxane chains and physically adsorbed polydimethylsiloxane. *Macromolecules* **2019**, *52*, 2863–2877. [CrossRef]
15. Ma, Y.; Lu, G.; Shao, C.; Li, X. Molecular dynamics simulation of hydrocarbon molecule adsorption on kaolinite (0 0 1) surface. *Fuel* **2019**, *237*, 989–1002. [CrossRef]
16. Xu, Y.; Liu, Y.-L.; Gao, S.; Jiang, Z.-W.; Su, D.; Liu, G.-S. Monolayer adsorption of dodecylamine surfactants at the mica/water interface. *Chem. Eng. Sci.* **2014**, *114*, 58–69. [CrossRef]
17. Silvestri, A.; Ataman, E.; Budi, A.; Stipp, S.; Gale, J.D.; Raiteri, P. Wetting Properties of the CO2–Water–Calcite System via Molecular Simulations: Shape and Size Effects. *Langmuir* **2019**, *35*, 16669–16678. [CrossRef]
18. Perilla, J.R.; Goh, B.C.; Cassidy, C.K.; Liu, B.; Bernardi, R.C.; Rudack, T.; Yu, H.; Wu, Z.; Schulten, K. Molecular dynamics simulations of large macromolecular complexes. *Curr. Opin. Struct. Biol.* **2015**, *31*, 64–74. [PubMed]
19. Ziebarth, J.; Wang, Y. Molecular dynamics simulations of DNA-polycation complex formation. *Biophys. J.* **2009**, *97*, 1971–1983. [CrossRef]
20. Tsai, J.-L.; Tu, J.-F. Characterizing mechanical properties of graphite using molecular dynamics simulation. *Mater. Des.* **2010**, *31*, 194–199.
21. Zhong, J.; Wang, P.; Zhang, Y.; Yan, Y.; Hu, S.; Zhang, J. Adsorption mechanism of oil components on water-wet mineral surface: A molecular dynamics simulation study. *Energy* **2013**, *59*, 295–300. [CrossRef]
22. Nikjoo, H.; O'Neill, P.; Wilson, W.; Goodhead, D. Computational approach for determining the spectrum of DNA damage induced by ionizing radiation. *Radiat. Res.* **2001**, *156*, (577–583. [CrossRef]
23. Du, H.; Miller, J. A molecular dynamics simulation study of water structure and adsorption states at talc surfaces. *Int. J. Miner. Process.* **2007**, *84*, 172–184. [CrossRef]
24. Wang, L.; Hu, Y.; Sun, W.; Sun, Y. Molecular dynamics simulation study of the interaction of mixed cationic/anionic surfactants with muscovite. *Appl. Surf. Sci.* **2015**, *327*, 364–370. [CrossRef]
25. Sepehrinia, K.; Mohammadi, A. Wettability alteration properties of fluorinated silica nanoparticles in liquid-loaded pores: An atomistic simulation. *Appl. Surf. Sci.* **2016**, *371*, 349–359. [CrossRef]

26. Zhao, J.; Yao, G.; Ramisetti, S.B.; Hammond, R.B.; Wen, D. Molecular dynamics investigation of substrate wettability alteration and oil transport in a calcite nanopore. *Fuel* **2019**, *239*, 1149–1161. [CrossRef]
27. Guo, J.; Zhang, L.; Liu, S.; Li, B. Effects of hydrophilic groups of nonionic surfactants on the wettability of lignite surface: Molecular dynamics simulation and experimental study. *Fuel* **2018**, *231*, 449–457. [CrossRef]
28. Naser, M.A.; Permadi, A.K.; Bae, W.; Ryoo, W.S.; Yunsun, P.; Dang, S.T.; Kim, S. Steam-Induced Wettability Alteration Through Contact Angle Measurement, a Case Study in X Field, Indonesia. *Proceedings of the SPE International Heavy Oil Conference and Exhibition*, Mangaf, Kuwait, 8–10 December 2014, OnePetro: Richardson, TX, USA, 2014.
29. Tipler, P.A.; Mosca, G. Physics for scientists and engineers. Macmillan: New York, NY, USA, 2007.
30. Yaseen, S.; Mansoori, G.A. Molecular dynamics studies of interaction between asphaltenes and solvents. *J. Pet. Sci. Eng.* **2017**, *156*, 118–124. [CrossRef]
31. Ahmadi, M.; Chen, Z. Comprehensive molecular scale modeling of anionic surfactant-asphaltene interactions. *Fuel* **2021**, *288*, 119729. [CrossRef]
32. BIOVIA, D.S. *Materials Studio*, 2020th ed.; Dassault Systèmes: San Diego, CA, USA, 2020.
33. Sun, H.; Ren, P.; Fried, J. The COMPASS force field: Parameterization and validation for phosphazenes. *Comput. Theor. Polym. Sci.* **1998**, *8*, 229–246. [CrossRef]
34. Sun, H. COMPASS: An ab initio force-field optimized for condensed-phase applications overview with details on alkane and benzene compounds. *J. Phys. Chem. B* **1998**, *102*, 7338–7364. [CrossRef]
35. Akkermans, R.L.; Spenley, N.A.; Robertson, S.H. COMPASS III: Automated fitting workflows and extension to ionic liquids. *Mol. Simul.* **2021**, *47*, 540–551.
36. Pacheco-Sánchez, J.; Zaragoza, I.; Martínez-Magadán, J. Asphaltene aggregation under vacuum at different temperatures by molecular dynamics. *Energy Fuels* **2003**, *17*, 1346–1355.
37. Xin, S.-M.; Liu, Q.-K.; Wang, K.; Chen, Y.; Yuan, P.-Q.; Cheng, Z.-M.; Yuan, W.-K. Solvation of asphaltenes in supercritical water: A molecular dynamics study. *Chem. Eng. Sci.* **2016**, *146*, 115–125. [CrossRef]
38. Bunte, S.W.; Sun, H. Molecular modeling of energetic materials: The parameterization and validation of nitrate esters in the COMPASS force field. *J. Phys. Chem. B* **2000**, *104*, 2477–2489. [CrossRef]
39. McQuaid, M.J.; Sun, H.; Rigby, D. Development and validation of COMPASS force field parameters for molecules with aliphatic azide chains. *J. Comput. Chem.* **2004**, *25*, 61–71. [CrossRef]
40. Ahmadi, M.; Hou, Q.; Wang, Y.; Chen, Z. Interfacial and molecular interactions between fractions of heavy oil and surfactants in porous media: Comprehensive review. *Adv. Colloid Interface Sci.* **2020**, 102242. [CrossRef] [PubMed]
41. Sun, H.; Jin, Z.; Yang, C.; Akkermans, R.L.; Robertson, S.H.; Spenley, N.A.; Miller, S.; Todd, S.M. COMPASS II: Extended coverage for polymer and drug-like molecule databases. *J. Mol. Model.* **2016**, *22*, (2), 47. [CrossRef]
42. Zhang, L.; LeBoeuf, E.J. A molecular dynamics study of natural organic matter: 1. Lignin, kerogen and soot. *Org. Geochem.* **2009**, *40*, 1132–1142.
43. Wu, G.; Zhu, X.; Ji, H.; Chen, D. Molecular modeling of interactions between heavy crude oil and the soil organic matter coated quartz surface. *Chemosphere* **2015**, *119*, 242–249. [CrossRef] [PubMed]
44. Xuefen, Z.; Guiwu, L.; Xiaoming, W.; Hong, Y. Molecular dynamics investigation into the adsorption of oil–water–surfactant mixture on quartz. *Appl. Surf. Sci.* **2009**, *255*, 6493–6498. [CrossRef]
45. Wu, G.; He, L.; Chen, D. Sorption and distribution of asphaltene, resin, aromatic and saturate fractions of heavy crude oil on quartz surface: Molecular dynamic simulation. *Chemosphere* **2013**, *92*, 1465–1471. [CrossRef]
46. Ahmadi, M.; Chen, Z. Molecular dynamics simulation of oil detachment from hydrophobic quartz surfaces during steam-surfactant Co-injection. *Energy* **2022**, *254*, 124434. [CrossRef]
47. Li, X.; Xue, Q.; Zhu, L.; Jin, Y.; Wu, T.; Guo, Q.; Zheng, H.; Lu, S. How to select an optimal surfactant molecule to speed up the oil-detachment from solid surface: A computational simulation. *Chem. Eng. Sci.* **2016**, *147*, 47–53. [CrossRef]
48. Liang, S.; Fang, T.; Xiong, W.; Ding, B.; Yan, Y.; Zhang, J. Oil detachment by modified nanoparticles: A molecular dynamics simulation study. *Comput. Mater. Sci.* **2019**, *170*, 109177. [CrossRef]
49. Gao, M.; Zhang, H.; Liang, S.; Zhao, S.; Guo, Z.; Hang, Y. Behavior of carbon nanotube-asphalt composites for dioxide capture based on molecular simulation. *Chem. Phys. Lett.* **2022**, *802*, 139756. [CrossRef]
50. Li, P.; Chan, M.; Froehlich, W. Steam injection pressure and the sagd ramp-up process. *J. Can. Pet. Technol.* **2009**, *48*, 36–41. [CrossRef]
51. Shin, H.; Polikar, M. New economic indicator to evaluate SAGD performance. *Proceedings of the SPE Western Regional Meeting*, Irvine, CA, USA, 30 March–1 April 2005, Society of Petroleum Engineers: London, UK, 2005.
52. Hautman, J.; Klein, M.L. Microscopic wetting phenomena. *Phys. Rev. Lett.* **1991**, *67*, 1763. [CrossRef]
53. Treiber, L.; Owens, W. A laboratory evaluation of the wettability of fifty oil-producing reservoirs. *Soc. Pet. Eng. J.* **1972**, *12*, 531–540. [CrossRef]
54. Li, B.; Liu, G.; Xing, X.; Chen, L.; Lu, X.; Teng, H.; Wang, J. Molecular dynamics simulation of CO_2 dissolution in heavy oil resin-asphaltene. *J. CO2 Util.* **2019**, *33*, 303–310. [CrossRef]

molecules

MDPI

Article

In Silico Screening of Metal–Organic Frameworks and Zeolites for He/N_2 Separation

Ivan V. Grenev [1,2,*] and Vladimir Yu. Gavrilov [2]

[1] Department of Physics, Novosibirsk State University, Pirogova Str. 1, Novosibirsk 630090, Russia
[2] Boreskov Institute of Catalysis, Ac. Lavrentiev Av. 5, Novosibirsk 630090, Russia
* Correspondence: greneviv@catalysis.ru

Abstract: In silico screening of 10,143 metal–organic frameworks (MOFs) and 218 all-silica zeolites for adsorption-based and membrane-based He and N_2 separation was performed. As a result of geometry-based prescreening, structures having zero accessible surface area (ASA) and pore limiting diameter (PLD) less than 3.75 Å were eliminated. So, both gases can be adsorbed and pass-through MOF and zeolite pores. The Grand canonical Monte Carlo (GCMC) and equilibrium molecular dynamics (EMD) methods were used to estimate the Henry's constants and self-diffusion coefficients at infinite dilution conditions, as well as the adsorption capacity of an equimolar mixture of helium and nitrogen at various pressures. Based on the obtained results, adsorption, diffusion and membrane selectivities as well as membrane permeabilities were calculated. The separation potential of zeolites and MOFs was evaluated in the vacuum and pressure swing adsorption processes. In the case of membrane-based separation, we focused on the screening of nitrogen-selective membranes. MOFs were demonstrated to be more efficient than zeolites for both adsorption-based and membrane-based separation. The analysis of structure–performance relationships for using these materials for adsorption-based and membrane-based separation of He and N_2 made it possible to determine the ranges of structural parameters, such as pore-limiting diameter, largest cavity diameter, surface area, porosity, accessible surface area and pore volume corresponding to the most promising MOFs for each separation model discussed in this study. The top 10 most promising MOFs were determined for membrane-based, vacuum swing adsorption and pressure swing adsorption separation methods. The effect of the electrostatic interaction between the quadrupole moment of nitrogen molecules and MOF atoms on the main adsorption and diffusion characteristics was studied. The obtained results can be used as a guide for selection of frameworks for He/N_2 separation.

Keywords: helium recovery; adsorption-based separation; membrane-based separation; metal–organic frameworks; zeolites; in silico screening; molecular simulation

1. Introduction

Helium is widely used in aviation, space and electronic industries, science and medicine due to its unique physical properties [1–3]. Today helium is mostly recovered from natural gas by the cryogenic distillation method. The traditional process used for helium separation [3–5] consists of the following stages: preliminary purification (separation from C_{2+} hydrocarbons, water, H_2S and CO_2), cryogenic separation of methane followed by cryogenic separation of nitrogen in a nitrogen rejection unit (NRU) yielding feedstock with helium concentration 1–3%, separation of the N_2/He mixture in a helium recovery unit yielding 50–70% crude helium vapor stream, purification of crude helium in a helium upgrade unit producing outcoming gas with helium concentration about 90%, deep purification in a helium purification unit producing helium with 99.995% purity and liquefaction of commodity helium. Membrane-based separation, vacuum swing adsorption (VSA) and pressure swing adsorption (PSA) processes are alternatives for the cryogenic distillation method [3,6,7]. Specific feature of these methods is that they can be used both as independent technological processes for helium recovery directly from the natural gas or as intermediate stages at existing factories using cryogenic distillation. According to the existing estimates [8,9], adsorption-based and membrane-based technologies are most expedient to be used at the stages of crude helium separation from the N_2/He mixture and deep purification. The efficiency of these processes is mostly determined by the physical properties of sorbents and materials used to produce membranes. In the case of the swing adsorption separation method, the sorbent must have high adsorption selectivity and working capacity. In the case of membrane-based separation, the material should possess high membrane selectivity and permeability [10–13]. Therefore, the search and development of novel efficient materials for both membrane-based and adsorption-based separation of the N_2/He mixture is an important problem.

Due to the diversity of the framework types, varying cation composition, high adsorption capacity, thermal stability and relatively low cost, zeolites are widely used in industry as sorbents in adsorption-based gas separation processes. Commercial zeolites potentially applicable for N_2/He separation are ZSM-5, HISIV 3000 (UOP), 5A (UOP, Sigma, St. Louis, MO, USA) and 13X (UOP) [3]. It is very difficult to estimate the adsorption selectivity of porous materials in separation of He and N_2 under ambient conditions because helium is poorly adsorbed on them. We are aware of only one paper [14] reporting an experimental estimation of helium sorption on zeolites 13X, 5A and 4A at room temperature and atmospheric pressure. Using these data as well data on nitrogen adsorption on these materials [15], it is possible to estimate their adsorption selectivity as 221, 458 and 54 for zeolites 13X, 5A and 4A (Table S1), respectively. For membrane-based separation, zeolites with framework types of DDR, MFI, STT and CHA are used. High membrane selectivity to helium can be achieved using porous frameworks with effective pore diameter close to the kinetic diameter of the nitrogen molecule. Polycrystalline membranes always have defects in the crystallite packing resulting in the loss of selectivity and increase of permeance. For example, a membrane based on zeolite STT [16] having strong size-exclusion effect for nitrogen demonstrated the He/N_2 selectivity equal to 11 at the average thickness of 2 μm and 59 at the average thickness of 6.6 μm. Zeolite 4A has weaker size-exclusion effect resulting in lower membrane He/N_2 selectivity ranging from 1.4 to 3.7 [17–20]. Membranes from zeolite DDR have selectivity close to 3 [21,22]. At low temperatures, zeolites membranes are nitrogen selective, since adsorption selectivity dominates over diffusion selectivity. The dependence of the membrane selectivity on temperature for ultrathin MFI membranes was evaluated in a wide temperature range by Yu et al. [23]. It was shown that the N_2/He membrane selectivity reaches 52 at 174 K for equimolar CH_4/N_2/He mixture at 3 bar feed pressure and 0.2 bar permeate pressure.

Metal–organic frameworks form a promising class of crystalline porous materials that can be used in the adsorption-based and membrane-based gas separation. The great variety of available organic and inorganic building blocks makes it possible to tune their adsorption properties, surface area, pore sizes and pore volume. Compared to zeolites, in the literature there are not as many experimental data on the selectivity and permeance of MOF membranes for the He/N_2 separation. MOF that is most widely used for synthesis of membranes is ZIF-8. According to the literature data [24–26], polycrystalline membranes made of ZIF-8 with the thickness varying from 17.5 to 80 μm are characterized by the He/N_2 selectivity in the range of 4.22–5.45 and He permeability between 3137 and 11354 Barrer. ZIF-8 single crystals with the size of 500 μm were used as membranes [27]. The He/N_2 selectivity of two studied single-crystal membranes was equal to 74.5 and 77.7 with the He permeability of 1935 and 2309 Barrer. Note that ZIF-8 has channels with an aperture of 3.4 Å, which is smaller than the kinetic diameter of the nitrogen molecule. The N_2 diffusion in the framework is possible due to the structural flexibility of ZIF-8. Hindered N_2 diffusion is reflected in low N_2 permeability equal to 26.0 and 29.7 Barrer for two different single-crystal membranes. Another studied MOF is HKUST-1. A polycrystalline membrane made of this material with the thickness of 40 μm was prepared on α-Al_2O_3 support [28]. Such membrane had the He/N_2 selectivity of 3.7. In the literature, there are also data on the He/N_2 selectivity of a membrane made of

IRMOF-1, which was equal to 2.4 [29]. Another representative of the IRMOF family IRMOF-3 demonstrated the selectivity of 2.5 [29]. In addition, the He/N_2 selectivity of membranes made of MIL-53 [30] and $[Cu_2(bza)_4(pyz)]_n$ [31] was reported to be equal to 2.4 and 3.9, respectively.

If the experimental measurement of the selectivity, adsorption capacity and permeability is not possible, simulation of adsorption and diffusion characteristics can give valuable information for estimation of the suitability of different materials for the He/N_2 separation. The development of zeolite database IZA [32] and MOF databases CSD MOF [33] and CoRE MOF [34] containing the unit cell parameters and atomic coordinates initiated in silico screening of efficient materials for adsorption-based [35–39] and membrane-based [40–42] separation. For example, Zarabadi-Poor et al. carried out in silico screening of almost 500 MOF structures from the DFT-optimized CoRE MOFs database [43] for helium separation from natural gas via N_2/He separation [44]. The API (Adsorbent Performance Indicator) metric [45] and membrane selectivity were used to determine the top performing MOF structures in the case of adsorption-based and membrane-based separation, respectively. We used the results obtained in this work as a reference for comparison. Most of the studies are focused on the screening of materials for membrane-based helium recovery from natural gas [10,46]. Kadioglu et al. screened 139 MOF structures for He/CH_4 membrane-based separation [47]. It was shown that the top performing MOF structures had a pore limiting diameter in the range of 3.8–4 Å. Qiao and coworkers predicted gas permeability and selectivity of MOFs for membrane-based separation of He/CH_4 and He/N_2 at infinite dilution condition [48]. Based on the obtained results, the Top 5 MOF structures for He/N_2 separation show selectivity in the range of 3.02 to 3.53 and permeability in the range of 8.72×10^2 to 8.72×10^3 Barrer. Daglar et al. studied incorporation of MOF fillers into polymers to obtain mixed matrix membranes (MMM) for 11 different gas separations processes [49]. It was shown that all MOF fillers improve He permeability with slight changes in He/N_2 membrane selectivity. Helium permeability and He/N_2 selectivity of such MMM varied from 39.2 to 1.13×10^4 Barrers and from 0.8 to 622, respectively. Gas permeabilities and selectivities for covalent organic frameworks (COFs) from the CURATED COF database [50] and COF/polymer MMMs for helium separation were predicted in two studies by different groups (Aydin et al. [51] and Feng et al. [52]). It was shown that COFs had a linear correlation between He permeability and He/N_2 selectivity. For He/N_2 separation, the selectivity of COF structures varies from 5.8×10^{-2} to 8.7. In both studies it was shown that the addition of COF fillers improved the He permeability of MMMs without significantly changing the He/N_2 selectivity. Based on the results of these studies, we think that MOF/COF membranes have low He selectivity at room temperature, while polymer membranes with high He/N_2 selectivity have low He permeability. Thus, two strategies are possible for creating efficient membranes for the He/N_2 separation: incorporation of porous fillers into the polymers to obtain MMMs with high He/N_2 selectivity and He permeability, or the use of nitrogen-selective porous medium membranes. Since the first approach has been extensively studied in the literature, we have focused on the screening of nitrogen-selective membranes. The goal of this study was to perform screening of zeolites and MOFs for search of relationships between their structural parameters and performance characteristics for adsorption-based and membrane-based He/N_2 separation.

2. Results and Discussion

2.1. Adsorption-Based Gas Separation

To test the force field models used in this study, calculated N_2 adsorption isotherms for a number of well-known MOFs at 295–298 K and pressure up to 20 bar were compared with the corresponding literature data [53–59]. Figure 1 (left) demonstrates that the suggested force field model can be used to predict the nitrogen adsorption in various MOFs with good precision. Several force fields were used to simulate the nitrogen adsorption in zeolites: Dreiding [60], TraPPE [61] and the force field developed by Vujić and Lyubartsev [62]. For each force field model, N_2 uptake values at 303 K and 1 bar pressure were calculated for five different pure silica frameworks, and the obtained results were compared with the corresponding experimental data [63] (Figure S1). The model by Vujić and Lyubartsev was used hereafter because it predicts the experimental data with the best precision. Figure 1 (right) demonstrates that this force field makes it possible to predict nitrogen adsorption isotherms with good precision both on pure silica zeolites and on several aluminophosphates [64,65]. The experimental conditions used to measure the nitrogen adsorption isotherms and the corresponding references are reported in Tables S2 and S3. Testing the used force field models for helium adsorption simulation is very difficult because the value measured in the adsorption experiment is excess adsorption. However, before the experiment, dead space is measured using the same helium. So, it is not possible to measure the helium adsorption isotherm by traditional methods.

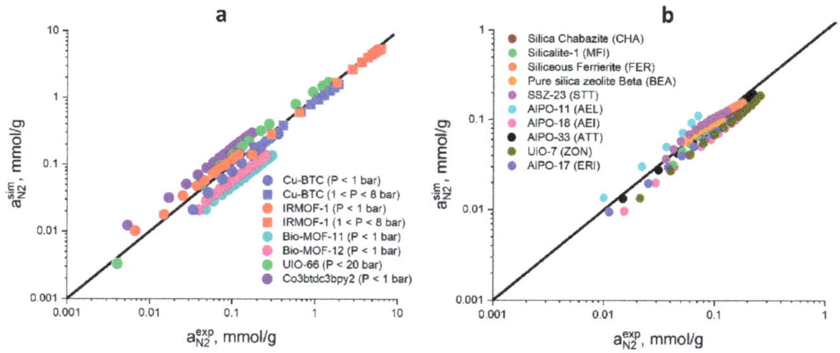

Figure 1. Comparison of experimental and molecular simulation data for N_2 adsorption in MOFs (**a**) and zeolites (**b**).

The performed screening of MOFs demonstrated that Henry's constants for He adsorption are in the range from 1.25×10^{-8} to 6.89×10^{-6} mol/kg/Pa, and for N_2 adsorption they vary from 7.08×10^{-8} to 1.21×10^{-3} mol/kg/Pa. The N_2/He adsorption selectivity at infinite dilution exceeds one for all the studied frameworks reaching a maximum value of 13,829. The highest Henry's constants for N_2 adsorption (Figure S4) and nitrogen adsorption selectivity (Figure S5) were observed at PLD about 4.5 Å and LCD about 6 Å. This result can be explained by the fact that the intermolecular interaction potentials overlap in small cavities leading to an increase in the heats of adsorption and Henry's constants. The analysis of the relationship between structural parameters and Henry's constants for He adsorption (Figure S4) revealed a trend to the K^0_{He} growth with an increase in accessible surface area, porosity and accessible pore volume. Meanwhile, a trend to the K^0_{He} decrease with an increase in the framework density was observed. In the case of K^0_{N2}, these trends were less evident due to the presence of an additional contribution from electrostatic interaction between the quadrupole moment of the nitrogen molecule and partial atomic charges of the MOF atoms. The analysis of the correlation between the structural parameters and N_2 adsorption selectivity (Figure S5) showed a trend of the decrease in $S^0_{ads,N2/He}$ with an increase in accessible surface area and porosity, and a decrease in density. Similar relationships between the structural parameters and adsorption selectivity to nitrogen were also observed for zeolites (Figure S5). The highest adsorption selectivity to nitrogen calculated at infinite dilution for zeolites was as high as 26. A large variety of the MOFs chemical composition leads to a greater variety of their structural parameters compared to zeolites (Figure S3). As a result, the selectivity of the most promising MOFs in adsorption-based separation of nitrogen–helium mixtures exceeds the best of the zeolite frameworks by several orders of magnitude.

To move from the estimation of ideal adsorption properties of MOFs and zeolites at infinite dilution to gas mixtures, adsorption of an equimolar mixture of helium and nitrogen was simulated at 0.01, 0.1, 0.3 and 1 Mpa and 298 K. The obtained results (Figure S6) demonstrate that at low adsorption selectivity $S^0_{ads,N_2/He}$ and $S^{mix}_{ads,N_2/He}$ are almost equal. However, when the adsorption selectivity increases, the $S^{mix}_{ads,N_2/He}/S^0_{ads,N_2/He}$ ratio becomes less than one.

When the pressure increases, the difference between the two selectivity values grows as well. So, the higher the pressure in the system, the greater the effect of competitive adsorption.

Two models of He/N_2 adsorption-based separation at room temperature were considered in this study. In the first model, corresponding to the conditions of vacuum swing adsorption, the adsorption pressure was equal to 0.1 Mpa and the desorption pressure was equal to 0.01 Mpa. In the second model, corresponding to conditions of pressure swing adsorption, the adsorption pressure was equal to 1 Mpa and the desorption pressure was equal to 0.1 Mpa. As regenerability tended to decrease with the APS increase, only frameworks with regenerability above 80% were considered. The highest APS values were observed for frameworks with high nitrogen ΔN and $S^{mix}_{ads,N_2/He}$. Top 50 best MOFs have APS > 57 mol/kg and APS > 70 mol/kg for the VSA and PSA gas separation models, respectively. (Figure 2a,b). In the case of PSA, the APS metric of the most promising MOFs is higher by more than an order of magnitude than that of the most efficient zeolites (Figure 2d). In the case of VSA, this difference reaches two orders of magnitude (Figure 2c). This result clearly demonstrates how promising MOFs are for adsorption-based separation of helium and nitrogen. In addition to the APS metric, sometimes a more complex API (Adsorbent Performance Indicator) metric [45], which additionally takes into account enthalpy of adsorption, is used in the literature:

$$API = \frac{\left(S^{mix}_{ads,N_2/He} - 1\right)^{A} \cdot \Delta N_{N2}{}^{B}}{\left|\Delta H_{ads,N2}\right|^{C}} \tag{1}$$

Here, $\Delta H_{ads,N2}$ is enthalpy of nitrogen adsorption; constants A, B and C are equal to 0.5, 2 and 1 [45]. As adsorption is an exothermal process, for large-scale industrial installations for separation based on vacuum swing adsorption, the use of an adsorbent with high adsorption enthalpy of results in an increase of the adsorber temperature leading to the decrease of the target component adsorption. On the other hand, heat is absorbed during the adsorbent regeneration leading to the decrease in the adsorber temperature, which makes the adsorbent regeneration more difficult. So, the higher the enthalpy of adsorption, the greater the difference of regenerability from the ideal model value. Therefore, an efficient adsorbent should have high adsorption capacity, high selectivity and low enthalpy of adsorption. In this study the API metric was calculated for all the studied MOFs and zeolites with regenerability above 80%. Figure S7 demonstrates a linear correlation between API and APS for both VSA and PSA. So, the use of both metrics leads to the same set of the most promising frameworks.

In this study, structure–adsorption performance relationships were investigated. Two databases with Top 50 MOFs based on the APS metric were constructed for the VSA and PSA separation processes. Then, smoothed probability density distributions (PDF) were built for several structural parameters for the database of all studied MOFs and for the databases of Top 50 MOFs for the VSA and PSA separation processes. These structural parameters included pore limiting diameter, largest cavity diameter, accessible surface area, accessible pore volume, density and porosity. The range of effective structural parameters was determined using a criterion that this range included more than 90% of Top 50 MOFs. The importance of each structural parameter was estimated by comparing PDFs for all studied MOFs and Top 50 MOFs. If PDFs were about the same for both databases, this structural parameter was considered to have little effect on the adsorption performance. Meanwhile, if PDFs were very different and PDF for TOP 50 MOFs had a narrow distribution, it was possible to claim that certain range of optimal parameters existed. For instance, Figure 3 demonstrates that for the VSA separation process there is a narrow range 3.75 Å < PLD < 4.8 Å corresponding to more than 90% of Top 50 MOFs, and PDF for them is different from PDF for all MOFs. So, the following optimal structural parameters were determined for VSA (Figure S8): 3.75 Å < PLD < 4.8 Å, 4.4 Å < LCD < 6 Å, 100 m^2/g < ASA < 700 m^2/g, 0.02 cm^3/g < AV < 0.09 cm^3/g, 1300 kg/m^3 < density < 2500 kg/m^3, 0.04 < VF < 0.15. For PSA the optimal ranges are wider (Figure S9): 3.75 Å < PLD < 6.3 Å,

4.2 Å < LCD < 7.1 Å, 300 m^2/g < ASA < 1400 m^2/g, 0.03 cm^3/g < AV < 0.18 cm^3/g, 900 kg/m^3 < density < 2500 kg/m^3, 0.05 < VF < 0.2. To determine the chemical composition–adsorption performance relationships, the probabilities of finding certain metal atoms in the MOF structure were calculated for all the studied MOFs and the database of Top 50 MOFs for the VSA and PSA separation processes (Figure 4). Ga, Ru, V, Er, Gd, La, U and Ca were found to be the most suitable metals for the VSA separation model based on the probabilities of their presence among Top 50 MOFs and all studied MOFs. Meanwhile, the most widespread metal atoms Zn, Cu, Cd, Co and Mn (present in 54.7% of all MOFs) were scarcely present among Top 50 MOFs (less than 10%). In the case of the PSA model, U, Al, Er, Be and Mg were found to be the most suitable metals. Characteristics of the most promising MOFs (Top 10) for the VSA and PSA separation processes are reported in Tables S6 and S7. As the pressure increase leads to a much more significant selectivity decrease than the adsorption capacity increase, the best MOFs for VSA have higher APS than the most promising MOFs for PSA.

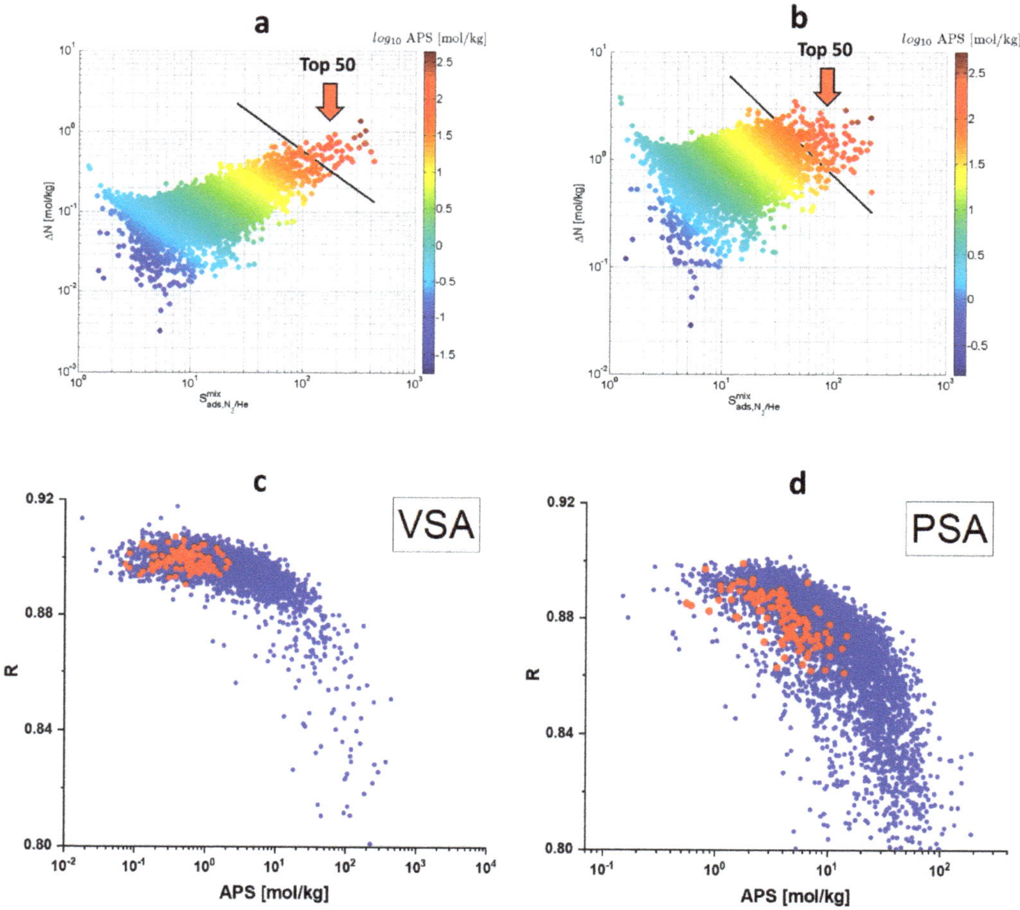

Figure 2. Dependence of the working capacity (ΔN) on adsorption selectivity ($S^{mix}_{ads,N_2/He}$) and APS in the case of vacuum swing adsorption (**a**) and pressure swing adsorption (**b**). Dependence of the regenerability (R) on APS for MOFs (blue) and zeolites (red) in the case of vacuum swing adsorption (**c**) and pressure swing adsorption (**d**).

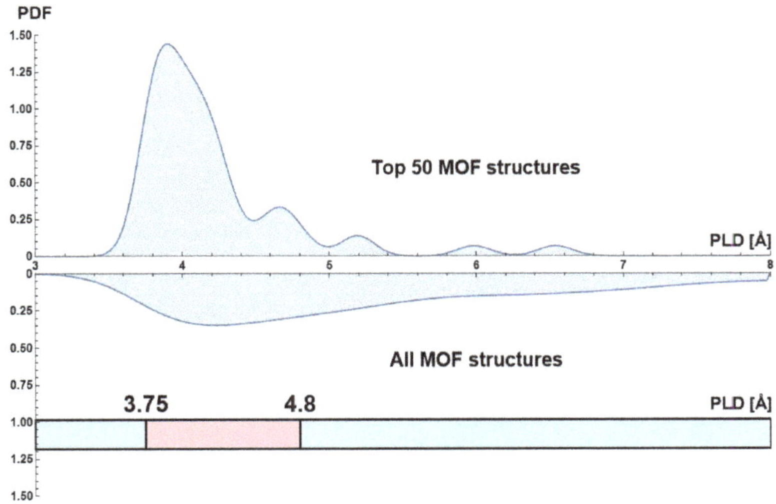

Figure 3. Comparison of the smoothed probability density distribution (PDF) of pore limiting diameter (PLD) for the best 50 MOFs and for all MOFs considered in this study for the VSA separation process. At the bottom of the figure is the PLD range that comprises over 90% of the Top 50 MOFs database.

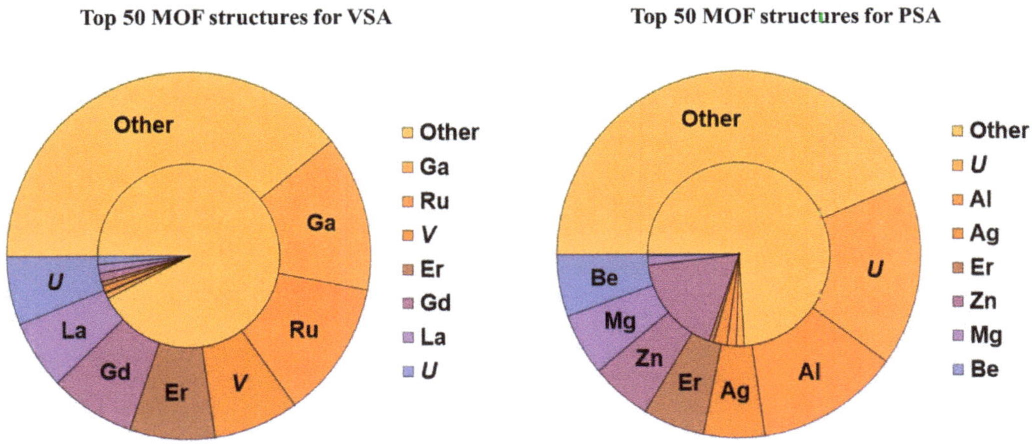

Figure 4. Chemical composition distribution of MOF structures by metal type. The outer pie chart corresponds to the Top 50 best MOFs, the inner one corresponds to all MOFs considered in this study.

It is interesting to compare the results of the MOF screening obtained in this study with earlier literature data. A total of 213 MOFs were studied for VSA gas separation with very similar operational conditions (equimolar mixture of helium and nitrogen, gas pressure during the adsorption cycle 1.2 bar, pressure during the adsorbent regeneration 0.1 bar) [44]. UVEXAV was found to be the most promising MOF with API = 680 and adsorption selectivity $S^{mix}_{ads,N_2/He}$ = 222.7. In our study we discovered 24 MOFs with superior API metric and 34 MOFs with higher nitrogen selectivity. Additionally, in the same earlier study [44] it was demonstrated that the electrostatic contribution for most MOFs was negligible. Despite the fact that a nitrogen molecule has a relatively a low quadrupole moment, we believe that it is important to consider the electrostatic interactions between nitrogen molecules and the framework atoms during the screening. To estimate their contributions, Henry's constants and nitrogen enthalpies of adsorption were calculated with the account of electrostatic interactions and without them (Figure S10). The account of electrostatic interactions results in the growth of the median value of Henry's constants and nitrogen adsorption enthalpies by 10.5% and 3.4%, respectively. Due to the great variety of the MOFs chemical composition, larger spread in Henry's constants and nitrogen adsorption enthalpies was observed for them in comparison with zeolites. In this study, it was found that the account of electrostatic interactions results in the increase of the Henry's constants by at least a factor of 1.5 for 15% of all considered MOFs. Meanwhile, the analysis of Top 50 MOFs for VSA and PSA revealed a similar increase in Henry's constants for 80% and 58% of MOFs, respectively. So, the contribution of electrostatic interactions between quadrupole moments of nitrogen molecules and the MOF atoms is significant for the most promising MOFs.

Since the chemical composition of both Top 50 MOFs includes a large number of different metal atoms with different atomic weights and partial charges, it can be assumed that metal atoms affect the adsorption performance primarily through the structure topology rather than through their contribution to the intermolecular interaction. Therefore, MOF screening for adsorption-based separation of N_2/He should be based primarily on the search for MOFs with optimal structural parameters. Thus, an ideal adsorbent should have uniformly narrow pores without pockets or cavities and a significant density in order to provide high nitrogen adsorption enthalpy and N_2/He selectivity. As a result, such an ideal structure will have low porosity, pore volume and accessible surface area.

2.2. Membrane-Based Gas Separation

To test the force fields used in this study for prediction of diffusion properties of MOFs and zeolites, calculated He and N_2 permeances in the temperature range of 298–301 K were compared with the corresponding literature data [22,26,29,30,66–71]. Figure S2 demonstrates that the models used in this study can be applied only for estimation of the membrane permeance. In addition to the lack of ideality in the used force field models, there are several additional reasons leading to deviation of the calculated permeance values from the experimental ones. First, most polycrystalline membranes based on MOFs and zeolites have low selectivity and high permeance due to the presence of defects both inside their framework and in the packing of crystals. The majority of membranes demonstrate selectivity close to that of the Knudsen diffusion model, which indicates that the size of pores between the crystallites can exceed 2 nm. Second, only permeability can be calculated by simulations. To connect this value with permeance, it is necessary to know the membrane thickness. Except for single-crystal membranes [27], the thickness of the MOF or zeolite layer on the support can be substantially varied resulting in deviation between the calculated and experimental values. Third, an ideal model imitating the adsorbate behavior at infinite dilution was used to calculate permeability in this study. Such a model does not take into account the effects of competitive adsorption and gas mixture composition on the diffusion properties. As helium is weakly adsorbed, and its concentration determined by GCMC is substantially lower than the N_2 concentration, it is necessary to consider much larger framework fragments than for simulation of adsorption to obtain correct values of self-diffusion coefficients. In turn, this leads to a significant increase in the simulation time. Therefore, it is rational to use the ideal model at the first screening stage with the following correction of the membrane permeability and selectivity for Top 10 frameworks using Equations (10) and (11).

At the first stage of screening for the most promising materials for membrane-based separation of He and N_2, diffusion coefficients were calculated at infinite dilution conditions. The analysis of the dependence of self-diffusion coefficients on the structural properties of the studied MOFs (Figure S12) demonstrates a natural trend towards an increase in self-diffusion coefficients with increasing pore size (PLD and LCD) and porosity. The greatest difference between $D^0_{N_2}$ and D^0_{He} was observed in the PLD range under 6 Å and LCD range under 8 Å. The diffusion selectivity $S^0_{dif\ He/N_2}$ in such small pores can reach several orders of magnitude (Figure S13a,b). This result can be explained by the difference in the kinetic diameters of helium and nitrogen molecules. Similar relationships between structural parameters and diffusion selectivity were also observed for zeolites (Figure S13). Similar to the adsorption-based separation, the screening demonstrated that great variety of MOFs makes it possible to find MOFs with much higher diffusion selectivity than that of zeolites (maximum $S^0_{dif\ He/N_2}$ for MOFs is equal to 1197 compared to 29 for zeolites).

The analysis of the screening results showed that the membrane selectivity $S^0_{mem\ He/N_2}$ varies from 2.4×10^{-3} to 4.6. Most of MOF membranes (79.9%) are N_2 selective. The upper part of the MOF "cloud" in Figure 5 with the membrane selectivity $S^0_{mem\ He/N_2} > 1$ is characterized by predomination of the diffusion selectivity $S^0_{dif\ He/N_2}$ over the adsorption selectivity $S^0_{ads\ He/N_2}$. The reverse ratio characterizes the bottom part of the MOF "cloud" in Figure 5. Helium permeability P^0_{He} is in the range from 2864 to 2.9×10^6 Barrer, whereas $P^0_{N_2}$ is in the range of 4034–1.4×10^7 Barrer. The highest permeabilities are observed for wide-pore frameworks. For examples, in Figure 5 one can see a "tail" consisting of 50 MOFs with $P^0_{He} > 5 \times 10^5$ Barrer. MOFs in this "tail" have PLD from 6.9 to 70.8 Å and LCD from 7 to 70.9 Å. Diffusion in these MOFs either follows the Knudsen diffusion model or is close to it. The diffusion selectivity in this diffusion mode is determined as $S^{\text{Knudsen}}_{dif,i/j} = \sqrt{M_j/M_i}$. For separation of He and N_2, this value is equal to 2.65.

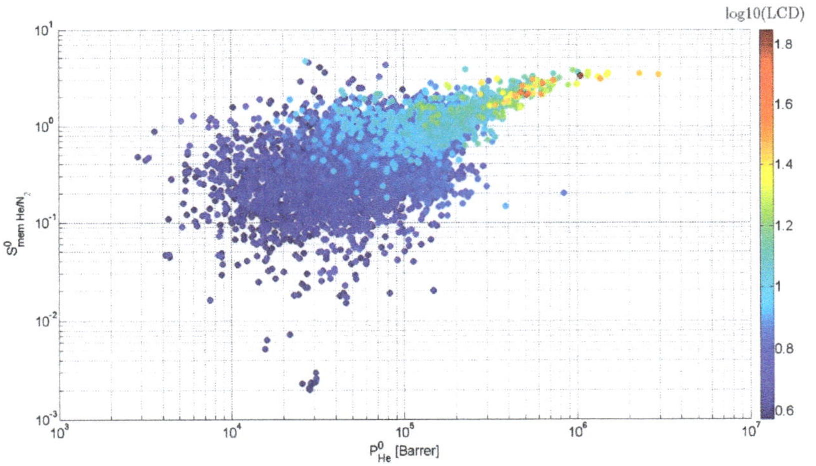

Figure 5. Dependence of membrane selectivity ($S^0_{mem\ He/N_2}$) on helium permeability (P^0_{He}) determined at infinite dilution and PLD.

In addition, the effect of electrostatic interaction on self-diffusion coefficients and nitrogen permeability as well as membrane selectivity to helium was studied (Figure S10). The account of electrostatic interaction results in a minor decrease in $D^0_{N_2\ (charged)}$ relative to $D^0_{N_2\ (non-charged)}$. As electrostatic interaction has a more significant effect on Henry's adsorption constants compared to self-diffusion coefficients, on the average the nitrogen permeability grows after the account of electrostatic interaction. As a result, the membrane selectivity to helium decreases. So, electrostatic interactions between the quadrupole moment of nitrogen molecules and the framework atoms should be taken into account during both adsorption-based and membrane-based screening.

The perspectives of using MOFs and zeolites as membrane materials for separation of He and N_2 can be estimated by comparing calculated selectivity and permeability with the upper bonds obtained in different years for polymeric membranes. Upper bond is an empirical relation $P_i = kS_{mem\ i/j}{}^n$ where constants k and n are determined from the selectivity vs. permeability graph for experimental data on He/N_2 separation over various polymeric membranes. Figure 6 presents upper bonds obtained by Robeson in 1991 [72] and 2008 [12] and by Wu et al. in 2019 [73]. The upper bonds for polymeric membranes demonstrate that the best MOFs and zeolites trail polymeric membranes in selectivity $S^0_{mem\ He/N_2}$ but have higher permeability P^0_{He}. Based on this fact, several variants of using MOFs and zeolites for separation of He and N_2 are possible: creation of materials from a combination of polymers and MOFs/zeolites or using MOF/zeolite materials selective to nitrogen. In the former case, combination of materials can result in the growth in permeability (in comparison with polymeric membranes) and increase in helium selectivity (in comparison with MOFs/zeolites). In several publications [49–51] it was shown that such an approach makes it possible to obtain more efficient materials for separation of different gas mixtures than using MOFs along as the membrane materials. However, due to the low membrane selectivity of MOFs and zeolites to helium, apparently, any other mesoporous materials with the membrane selectivity close to the Knudsen diffusion selectivity can be used for separation of helium and nitrogen. In the second case, if MOFs or zeolites selective to nitrogen are used as the membrane, the retentate will be enriched with helium and the permeate will be enriched with nitrogen. As using MOFs and zeolites as membrane materials is more promising in the second case, Top 50 frameworks most selective to nitrogen ($S^0_{mem\ N_2/He}$ > 22.6) were identified. Not a single zeolite was included in this list.

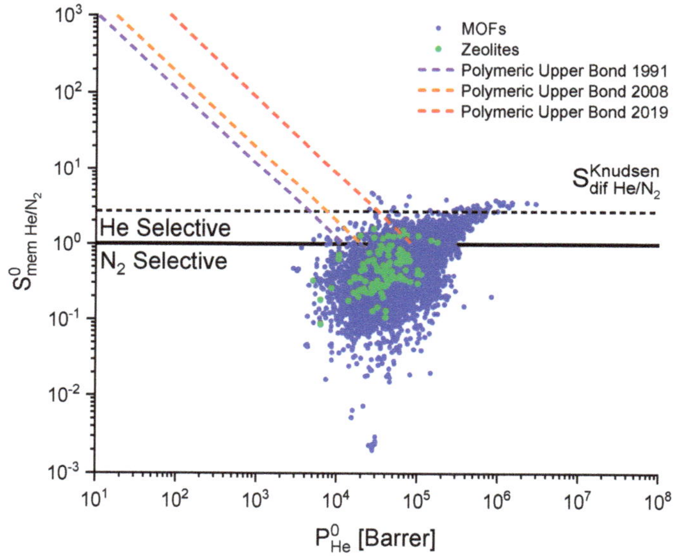

Figure 6. Dependence of membrane selectivity ($S^0_{mem\ He/N_2}$) on helium permeability (P^0_{He}) determined at infinite dilution. Lines correspond to the upper bonds for polymeric materials.

Structure–membrane separation performance relationships were also studied. Their analysis was performed using the same technique that was used for adsorption-based separation and was described above in Section 3.1. The following optimal structural parameters were determined for membrane-based separation (Figure S14): 3.75 Å < PLD < 4.65 Å, 4 Å < LCD < 5.8 Å, 100 m^2/g < ASA < 450 m^2/g, 0.01 cm^3/g < AV < 0.06 cm^3/g, 1100 kg/m^3 < density < 2400 kg/m^3, 0.02 < VF < 0.1. Note that the PDF functions for Top 50 MOFs and all the studied MOFs are substantially different, and the range of optimal structural parameters is much narrower than for adsorption-based separation. Similarly, chemical composition–membrane separation performance relationships were analyzed. Top 50 MOFs include 22 metals. It means that the membrane properties are primarily determined by its structural parameters rather than by chemical composition. Still, based on the ratio of probabilities of the presence among Top 50 MOFs and in all the studied MOFs, Ga and Al were found to be the most promising metals for membrane-based separation.

The characteristics of the most promising MOFs (Top 10) for membrane-based separation are reported in Table S8. All MOFs from Top 10 demonstrate high adsorption selectivity predominating over their diffusion selectivity and high N_2 permeability exceeding the He permeability by more than an order of magnitude. At the second stage of screening, these MOFs were used in simulation by GCMC and MD using an equimolar mixture of He and N_2 at 298 K and 3 bar, which corresponds to the conditions of real membrane-based separation of gases. The simulation results (Table S8) indicate that the switch to the equimolar mixture leads to simultaneous decrease of the nitrogen adsorption selectivity, He/N_2 diffusion selectivity and permeability. Nevertheless, the values of $S^0_{mem\ N_2/He}$ and $S^{mix}_{mem\ N_2/He}$ are very similar indicating that the suggested technique used for the initial screening makes it possible to predict promising frameworks while using minimum computer power. Earlier [44], based on the results of in silico screening of 500 MOFs, the highest membrane selectivity $S^0_{mem\ N_2/He}$ = 44.91 was found for LIFWOO. In this study, we determined 16 MOFs with even higher membrane selectivity. Thus, highly nitrogen-selective membranes are characterized by the predominance of the N_2/He adsorption selectivity over N_2/He diffusion selectivity. Therefore, as in the case of adsorption-based separation, an ideal membrane should have narrow uniform pores and a significant density. The membrane properties are primarily determined by its structural parameters rather than by chemical composition.

3. Computational Methods

3.1. MOF and Zeolite Databases

"Computation Ready, Experimental Metal−Organic Framework Database" (CoRE MOF 2019) [34] was used as the parent MOF database. At the first step only ordered frameworks where all solvent molecules were removed were selected from this database. The resulting database consisted of 10143 MOFs. Then, the Zeo++ algorithm [74] was used to calculate density, porosity (VF), accessible surface area (ASA), pore limiting diameter (PLD) and largest cavity diameter (LCD) for each framework. Additionally, Zeo++ was used to identify MOFs with pockets not accessible both for helium and nitrogen molecules. In the following computations, such pockets were blocked. The accessible pore volume (AV) was determined using RASPA-2.0 software package [75] with helium molecule used as a probe. Further, only MOFs with non-zero ASA and PLD > 3.75 Å were selected from the database. These conditions would allow both gases to penetrate through the membrane (the kinetic diameter is equal 3.64 Å for N_2 and 2.56 Å for He). So, the number of analyzed MOFs was shortened to 5156.

IZA database [32] implemented in iRASPA visualization package [76] was used as the parent database of all-silica zeolites and zeolite-like materials. Structural parameters were determined for all frameworks from this database using a procedure similar to the one used earlier for MOFs. After removing frameworks with PLD below 3.75 Å, the number of analyzed zeolite frameworks was reduced to 110.

3.2. Simulation Details

The adsorption and diffusion properties of the studied frameworks were simulated using equilibrium molecular dynamics (EMD) and Grand canonical Monte Carlo (GCMC) methods in the RASPA-2.0 package. The interactions were described by the sum of Lennard-Jones 6–12 (LJ) and Coulomb interaction potentials. The nitrogen molecule was simulated according to the TraPPE [77] force field as a dumbbell with a rigid bond between the atoms with the length of 1.1 Å. The LJ interaction parameters for each atom were ε_N/k_B = 36.0 K and σ_N = 3 31 Å. The N_2 quadrupole moment was described by three charges: two $-q$ charges where q = 0.482e that were located at the centers of the nitrogen atoms and $+2q$ charge located at the molecule center of masses. The helium molecule was simulated by a single-charge model [78] with parameters σ_{He} = 2.64 Å and ε_{He}/k_B = 10.9 K. The constants of the LJ 6–12 potential for atoms in MOFs were simulated according to the Dreiding [60] force field. If the necessary parameters were missing, the required data were retrieved from the UFF [79] force field. The force field developed by Vujić and Lyubartsev [62] was used for simulation of adsorption and diffusion in zeolites. Cross constants of the LJ interaction were calculated using the Lorentz-Berthelot mixing rule. The LJ interaction was cut at the distance R_{cutoff} = 12.8 Å. Its potential was shifted to zero starting from the distance 0.9 R_{cutoff}. Determination of partial atomic charges by REPEAT [80] or DDEC [81] methods that have high precision requires periodic DFT calculations for each studied structure. A detailed and systematic analysis of the effect of the choice of framework partial atomic charges on CO_2 adsorption in six different widely studied MOFs predicted by molecular simulations was performed in the study [82]. It was shown that the partial charges obtained by both DDEC and REPEAT methods yielded isotherms that were in good agreement with each other. In our previous studies, it was shown that partial charges obtained by DDEC and REPEAT methods lead to similar results for modeling hydrogen adsorption in SAPO-11 (zeolite-like material) at 77 K [83] and water adsorption in CAU-10-H (MOF) at 298 K [84]. Thus, it can be expected that the use of the DDEC or REPEAT methods for calculating atomic partial charges will lead to similar results in the case of modeling nitrogen adsorption in MOFs and zeolites. A reasonable alternative is to use pre-trained models obtained by machine learning based on the CoRE MOF DDEC [43,85] containing 2932 optimized structures with partial atomic charges calculated by the DDEC method. So, partial atomic charges of the framework atoms were determined using pre-trained Random Forest model in the PACMOF code [86]. Long-range Coulomb interactions were calculated using the Ewald summation technique. The size of the MOF and zeolite structure fragments was selected to ensure the minimum distance in each direction exceeded $2R_{cutoff}$. In each simulation, we assumed that the MOF or zeolite structure was rigid and did not contain any defects.

Henry's constants were calculated using the Widom particle insertion method. Henry's constants were calculated at infinite dilution conditions for 10^5 cycles at 298 K. The adsorption selectivity at infinite dilution $S^0_{ads,i/j}$ was determined as the ratio of Henry's constants K^0_i [10]:

$$S^0_{ads,i/j} = K^0_i / K^0_j \tag{2}$$

The adsorption selectivity for separation of gas mixtures $S^{mix}_{ads,i/j}$ was calculated for 0.01, 0.1, 0.3 and 1 MPa pressures at 298 K as follows [10]:

$$S^{mix}_{ads,i/j} = \frac{N_i / N_j}{x_i / x_j} \tag{3}$$

Here, N_i is the adsorption of the i-th component determined from GCMC, x_i is the molar fraction of the i-th component. For a gas mixture, the adsorption values N_i were simulated using the following GCMC moves: translation, rotation, insertion, deletion, reinsertion, identity exchange of He and N_2 molecules. The fugacity coefficient was calculated from the Peng–Robinson equation of state.

Adsorbent Performance Score (APS) used to estimate the adsorbent efficiency was calculated as follows [87]:

$$APS = \Delta N_{N2} S^{mix}_{ads,N2/He} \tag{4}$$

The sorbent regenerability was calculated as [10]:

$$R = \Delta N_{N2} / N_{ads,N2} \tag{5}$$

Self-diffusion coefficient was determined by equilibrium molecular dynamics from the root-mean-square particle displacement using the following formula [88]:

$$D_{self,i} = \lim_{t \to \infty} \frac{1}{2dt} \langle \frac{1}{N} \sum_{j=1}^{N} \left(\vec{r_j}(t) - \vec{r_j}(0) \right)^2 \rangle \tag{6}$$

Here, N is the number of molecules, $\vec{r_j}(t)$ is the position of the j-th particle at the moment t, d is the dimension of the system. Self-diffusion coefficient at infinite dilution D^0_i was simulated by positioning 30 adsorbate molecules in the MOF fragment with all interactions between the gas molecules switched off. The simulation was performed in the NVT ensemble (at constant number of particles, volume and temperature) using the Nosé–Hoover thermostat. These conditions simulate the properties of the adsorbate at infinite dilution. After the initial position of the adsorbate in the framework was generated, for the first 50 ps the system was subjected to equilibration before the data collection. The MD simulation time was 20 ns. The diffusion selectivity at infinite dilution $S^0_{dif,i/j}$ was calculated as follows [42]:

$$S^0_{dif,i/j} = D^0_i / D^0_j \quad (7)$$

Permeability P^0_i was determined using Equation (7) [42]:

$$P^0_i = D^0_i K^0_i \quad (8)$$

Membrane selectivity $S^0_{mem,i/j}$ was estimated as follows [42]:

$$S^0_{mem,i/j} = P^0_i / P^0_j \quad (9)$$

For selected Top 10 frameworks, self-diffusion coefficient $D_{self,i}$ was calculated for each component of the gas mixture at the adsorbate concentrations determined by GCMC. As helium is weakly adsorbed, and its concentration determined by GCMC is significantly lower than that of N_2, the framework fragment used in the EMD method was 27 times larger than during the initial screening at infinite dilution. Five independent EMD runs were performed to collect statistics. Diffusion selectivity for a mixture of gases $S^{mix}_{dif,i/j}$ was determined as the ratio of self-diffusion coefficients [42]:

$$S^{mix}_{dif,i/j} = \frac{D_{self,i}}{D_{self,j}} \quad (10)$$

Permeability of the i-th mixture component P^{mix}_i was calculated according to Equation (10) [42]:

$$P^{mix}_i = \frac{\varphi \cdot D_{self,i} \cdot c_i}{f_i} \quad (11)$$

Here, φ is the adsorbent porosity, c_i is the concentration of the i-th mixture component at the feed side of the membrane determined by GCMC, f_i is the fugacity of the i-th mixture component before the membrane. This model assumes vacuum permeating pressure. The membrane selectivity for the mixture of gases $S^{mix}_{mem,i/j}$ was determined as follows [42]:

$$S^{mix}_{mem,i/j} = P^{mix}_i / P^{mix}_j \quad (12)$$

4. Conclusions

In silico screening of 10143 MOFs and 218 all-silica zeolites for adsorption-based and membrane-based separation of He and N_2 was performed. GCMC and EMD methods were used to calculate Henry's constants, adsorption at different pressures and self-diffusion coefficients for He and N_2. These values were used to calculate major metrics, such as adsorption, diffusion and membrane selectivity, regenerability and permeability. Dependence of adsorption selectivity $S^{mix}_{ads,N_2/He}$ in an equimolar mixture of He and N_2 on the pressure in the system was studied. The effect of electrostatic interaction between the quadrupole moment of nitrogen molecules and framework atoms on the main adsorption and diffusion metrics was studied. MOFs were demonstrated to be more efficient than zeolites in both adsorption-based and membrane-based separation. Top 10 most promising MOFs for membrane-based, VSA and PSA separation methods were determined. The analysis of structure–adsorption and membrane performance relationships made it possible to determine the ranges of structural parameters, such as pore limiting diameter, largest cavity diameter, surface area, porosity, accessible surface area and pore volume, corresponding to the most promising MOFs for each separation model considered in this study. A similar analysis was performed to find out the optimal MOF chemical composition. The screening performed in this study can be called topological because the structural flexibility, possible presence of structural defects or modification of linkers with functional groups were not taken into account. Similarly, in the case of zeolites possible changes of the Si/Al ratio, variation of the cation composition or modification of their structure by isomorphous substitutions were not considered. Despite these limitations, one can expect that the ranges of optimal structural parameters and trends of adsorption and membranes metrics determined in this study will be correct even if all these factors are taken into account. The obtained results can be used as a guide for selection of frameworks for He/N_2 separation.

Supplementary Materials: The following supporting information can be downloaded at: https://www.mdpi.com/article/10.3390/molecules28010020/s1, Section S1: Comparison of molecular simulations and experimental data; Section S2: Structure—adsorption performance relationships; Section S3: Comparison of ideal and mixture adsorption selectivities; Section S4: Relationship between API and APS metrics; Section S5: Comparison of the structural properties PDF for the best 50 MOF structures for VSA and PSA with all MOF structures; Section S6: Characteristics of Top 10 MOFs for adsorption-based separation; Section S7: Effect of electrostatic interaction on He/N_2 separation; Section S8: Structure—diffusion performance relationships; Section S9: Comparison of the structural properties PDF for the best 50 MOF structures for membrane separation process with all MOF structures; Section S10: Chemical composition—membrane separation performance relationships; Section S11: Characteristics of Top 10 MOFs for membrane-based separation; References.

Author Contributions: Conceptualization, I.V.G. and V.Y.G.; Methodology, I.V.G. and V.Y.G.; Software, I.V.G.; Validation, I.V.G.; Investigation, I.V.G.; Data Curation, I.V.G.; Writing—Original Draft Preparation, I.V.G.; Writing—Review and Editing, I.V.G. and V.Y.G.; Visualization, I.V.G.; Supervision, V.Y.G.; Project Administration, I.V.G. All authors have read and agreed to the published version of the manuscript.

Funding: The reported study was supported by Russian Foundation for Basic Research, project number 19-33-60087.

Institutional Review Board Statement: Not applicable.

Informed Consent Statement: Not applicable.

Data Availability Statement: Data available on request.

Acknowledgments: The authors are grateful to the Siberian Supercomputer Center and Supercomputing Center of the Novosibirsk State University for providing supercomputer equipment.

Conflicts of Interest: The authors declare no conflict of interest.

References

1. Berganza, C.J.; Zhang, J.H. The Role of Helium Gas in Medicine. *Med. Gas Res.* **2013**, *3*, 18. [CrossRef] [PubMed]
2. Mahesh, M.; Barker, P.B. The MRI Helium Crisis: Past and Future. *J. Am. Coll. Radiol.* **2016**, *13*, 1536–1537. [CrossRef] [PubMed]
3. Rufford, T.E.; Chan, K.I.; Huang, S.H.; May, E.F. A Review of Conventional and Emerging Process Technologies for the Recovery of Helium from Natural Gas. *Adsorpt. Sci. Technol.* **2014**, *32*, 49–72. [CrossRef]
4. Dai, Z.; Deng, J.; He, X.; Scholes, C.A.; Jiang, X.; Wang, B.; Guo, H.; Ma, Y.; Deng, L. Helium Separation Using Membrane Technology: Recent Advances and Perspectives. *Sep. Purif. Technol.* **2021**, *274*, 119044. [CrossRef]
5. Sunarso, J.; Hashim, S.S.; Lin, Y.S.; Liu, S.M. Membranes for Helium Recovery: An Overview on the Context, Materials and Future Directions. *Sep. Purif. Technol.* **2017**, *176*, 335–383. [CrossRef]
6. Scholes, C.A.; Ghosh, U. Helium Separation through Polymeric Membranes: Selectivity Targets. *J. Membr. Sci.* **2016**, *520*, 221–230. [CrossRef]
7. Scholes, C.; Ghosh, U. Review of Membranes for Helium Separation and Purification. *Membranes* **2017**, *7*, 9. [CrossRef]
8. Alders, M.; Winterhalder, D.; Wessling, M. Helium Recovery Using Membrane Processes. *Sep. Purif. Technol.* **2017**, *189*, 433–440. [CrossRef]
9. Scholes, C.A. Helium Recovery through Inorganic Membranes Incorporated with a Nitrogen Rejection Unit. *Ind. Eng. Chem. Res.* **2018**, *57*, 3792–3799. [CrossRef]
10. Daglar, H.; Keskin, S. Recent Advances, Opportunities, and Challenges in High-Throughput Computational Screening of MOFs for Gas Separations. *Coord. Chem. Rev.* **2020**, *422*, 213470. [CrossRef]

11. Yang, S.; Yin, K.; Wu, J.; Wu, Z.; Chu, D.; He, J.; Duan, J.-A. Ultrafast Nano-Structuring of Superwetting Ti Foam with Robust Antifouling and Stability towards Efficient Oil-in-Water Emulsion Separation. *Nanoscale* **2019**, *11*, 17607–17614. [CrossRef] [PubMed]
12. Robeson, L.M. The Upper Bound Revisited. *J. Membr. Sci.* **2008**, *320*, 390–400. [CrossRef]
13. Yin, K.; Chu, D.; Dong, X.; Wang, C.; Duan, J.-A.; He, J. Femtosecond Laser Induced Robust Periodic Nanoripple Structured Mesh for Highly Efficient Oil–Water Separation. *Nanoscale* **2017**, *9*, 14229–14235. [CrossRef] [PubMed]
14. Malbrunot, P.; Vidal, D.; Vermesse, J.; Chahine, R.; Bose, T.K. Adsorbent Helium Density Measurement and Its Effect on Adsorption Isotherms at High Pressure. *Langmuir* **1997**, *13*, 539–544. [CrossRef]
15. Khoramzadeh, E.; Mofarahi, M.; Lee, C.-H. Equilibrium Adsorption Study of CO_2 and N_2 on Synthesized Zeolites 13X, 4A, 5A, and Beta. *J. Chem. Eng. Data* **2019**, *64*, 5648–5664. [CrossRef]
16. Gong, C.; Peng, X.; Zhu, M.; Zhou, T.; You, L.; Ren, S.; Wang, X.; Gu, X. Synthesis and Performance of STT Zeolite Membranes for He/N_2 and He/CH_4 Separation. *Sep. Purif. Technol.* **2022**, *301*, 121927. [CrossRef]
17. Xu, X.; Yang, W.; Liu, J.; Lin, L.; Stroh, N.; Brunner, H. Synthesis of NaA Zeolite Membrane on a Ceramic Hollow Fiber. *J. Membr. Sci.* **2004**, *229*, 81–85. [CrossRef]
18. Aoki, K.; Kusakabe, K.; Morooka, S. Separation of Gases with an A-Type Zeolite Membrane. *Ind. Eng. Chem. Res.* **2000**, *39*, 2245–2251. [CrossRef]
19. Aoki, K.; Kusakabe, K.; Morooka, S. Gas Permeation Properties of A-Type Zeolite Membrane Formed on Porous Substrate by Hydrothermal Synthesis. *J. Membr. Sci.* **1998**, *141*, 197–205. [CrossRef]
20. Guan, G.; Kusakabe, K.; Morooka, S. Gas Permeation Properties of Ion-Exchanged LTA-Type Zeolite Membranes. *Sep. Sci. Technol.* **2001**, *36*, 2233–2245. [CrossRef]
21. Himeno, S.; Tomita, T.; Suzuki, K.; Nakayama, K.; Yajima, K.; Yoshida, S. Synthesis and Permeation Properties of a DDR-Type Zeolite Membrane for Separation of CO_2/CH_4 Gaseous Mixtures. *Ind. Eng. Chem. Res.* **2007**, *46*, 6989–6997. [CrossRef]
22. Tomita, T.; Nakayama, K.; Sakai, H. Gas Separation Characteristics of DDR Type Zeolite Membrane. *Microporous Mesoporous Mater.* **2004**, *68*, 71–75. [CrossRef]
23. Yu, L.; Mayne, B.; Nobandegani, M.S.; Grekou, T.; Hedlund, J. Recovery of Helium from Natural Gas Using MFI Membranes. *J. Membr. Sci.* **2022**, *644*, 120113. [CrossRef]
24. Drobek, M.; Bechelany, M.; Vallicari, C.; Abou Chaaya, A.; Charmette, C.; Salvador-Levehang, C.; Miele, P.; Julbe, A. An Innovative Approach for the Preparation of Confined ZIF-8 Membranes by Conversion of ZnO ALD Layers. *J. Membr. Sci.* **2015**, *475*, 39–46. [CrossRef]
25. Hara, N.; Yoshimune, M.; Negishi, H.; Haraya, K.; Hara, S.; Yamaguchi, T. ZIF-8 Membranes Prepared at Miscible and Immiscible Liquid–Liquid Interfaces. *Microporous Mesoporous Mater.* **2015**, *206*, 75–80. [CrossRef]
26. Hara, N.; Yoshimune, M.; Negishi, H.; Haraya, K.; Hara, S.; Yamaguchi, T. Diffusive Separation of Propylene/Propane with ZIF-8 Membranes. *J. Membr. Sci.* **2014**, *450*, 215–223. [CrossRef]
27. Chen, C.; Ozcan, A.; Yazaydin, A.O.; Ladewig, B.P. Gas Permeation through Single-Crystal ZIF-8 Membranes. *J. Membr. Sci.* **2019**, *575*, 209–216. [CrossRef]
28. Hara, N.; Yoshimune, M.; Negishi, H.; Haraya, K.; Hara, S.; Yamaguchi, T. Metal–Organic Framework Membranes with Layered Structure Prepared within the Porous Support. *RSC Adv.* **2013**, *3*, 14233. [CrossRef]
29. Yoo, Y.; Varela-Guerrero, V.; Jeong, H.-K. Isoreticular Metal−Organic Frameworks and Their Membranes with Enhanced Crack Resistance and Moisture Stability by Surfactant-Assisted Drying. *Langmuir* **2011**, *27*, 2652–2657. [CrossRef]
30. Zhang, Y.; Gao, Q.; Lin, Z.; Zhang, T.; Xu, J.; Tan, Y.; Tian, W.; Jiang, L. Constructing Free Standing Metal Organic Framework MIL-53 Membrane Based on Anodized Aluminum Oxide Precursor. *Sci. Rep.* **2015**, *4*, 4947. [CrossRef] [PubMed]
31. Huang, A.; Chen, Y.; Liu, Q.; Wang, N.; Jiang, J.; Caro, J. Synthesis of Highly Hydrophobic and Permselective Metal–Organic Framework Zn(BDC)(TED)0.5 Membranes for H_2/CO_2 Separation. *J. Membr. Sci.* **2014**, *454*, 126–132. [CrossRef]
32. Database of Zeolite Structures. Available online: http://www.iza-structure.org/databases/ (accessed on 3 November 2022).
33. Moghadam, P.Z.; Li, A.; Wiggin, S.B.; Tao, A.; Maloney, A.G.P.; Wood, P.A.; Ward, S.C.; Fairen-Jimenez, D. Development of a Cambridge Structural Database Subset: A Collection of Metal–Organic Frameworks for Past, Present, and Future. *Chem. Mater.* **2017**, *29*, 2618–2625. [CrossRef]
34. Chung, Y.G.; Haldoupis, E.; Bucior, B.J.; Haranczyk, M.; Lee, S.; Zhang, H.; Vogiatzis, K.D.; Milisavljevic, M.; Ling, S.; Camp, J.S.; et al. Advances, Updates, and Analytics for the Computation-Ready, Experimental Metal–Organic Framework Database: CoRE MOF 2019. *J. Chem. Eng. Data* **2019**, *64*, 5985–5998. [CrossRef]
35. Altintas, C.; Avci, G.; Daglar, H.; Nemati Vesali Azar, A.; Erucar, I.; Velioglu, S.; Keskin, S. An Extensive Comparative Analysis of Two MOF Databases: High-Throughput Screening of Computation-Ready MOFs for CH_4 and H_2 Adsorption. *J. Mater. Chem. A* **2019**, *7*, 9593–9608. [CrossRef]
36. Solanki, V.A.; Borah, B. High-Throughput Computational Screening of 12,351 Real Metal–Organic Framework Structures for Separation of Hexane Isomers: A Quest for a Yet Better Adsorbent. *J. Phys. Chem. C* **2020**, *124*, 4582–4594. [CrossRef]
37. Basdogan, Y.; Sezginel, K.B.; Keskin, S. Identifying Highly Selective Metal Organic Frameworks for CH_4/H_2 Separations Using Computational Tools. *Ind. Eng. Chem. Res.* **2015**, *54*, 8479–8491. [CrossRef]
38. Fischer, M. Computational Evaluation of Aluminophosphate Zeotypes for CO_2/N_2 Separation. *Phys. Chem. Chem. Phys.* **2017**, *19*, 22801–22812. [CrossRef]

39. Erdős, M.; Geerdink, D.F.; Martin-Calvo, A.; Pidko, E.A.; van den Broeke, L.J.P.; Calero, S.; Vlugt, T.J.H.; Moultos, O.A. In Silico Screening of Zeolites for High-Pressure Hydrogen Drying. *ACS Appl. Mater. Interfaces* **2021**, *13*, 8383–8394. [CrossRef]
40. Daglar, H.; Keskin, S. High-Throughput Screening of Metal Organic Frameworks as Fillers in Mixed Matrix Membranes for Flue Gas Separation. *Adv. Theory Simul.* **2019**, *2*, 1900109. [CrossRef]
41. Budhathoki, S.; Ajayi, O.; Steckel, J.A.; Wilmer, C.E. High-Throughput Computational Prediction of the Cost of Carbon Capture Using Mixed Matrix Membranes. *Energy Environ. Sci.* **2019**, *12*, 1255–1264. [CrossRef]
42. Krishna, R.; van Baten, J.M. In Silico Screening of Zeolite Membranes for CO_2 Capture. *J. Membr. Sci.* **2010**, *360*, 323–333. [CrossRef]
43. Nazarian, D.; Camp, J.S.; Chung, Y.G.; Snurr, R.Q.; Sholl, D.S. Large-Scale Refinement of Metal−Organic Framework Structures Using Density Functional Theory. *Chem. Mater.* **2017**, *29*, 2521–2528. [CrossRef]
44. Zarabadi-Poor, P.; Marek, R. Metal–Organic Frameworks for Helium Recovery from Natural Gas via N_2/He Separation: A Computational Screening. *J. Phys. Chem. C* **2019**, *123*, 3469–3475. [CrossRef]
45. Wiersum, A.D.; Chang, J.-S.; Serre, C.; Llewellyn, P.L. An Adsorbent Performance Indicator as a First Step Evaluation of Novel Sorbents for Gas Separations: Application to Metal–Organic Frameworks. *Langmuir* **2013**, *29*, 3301–3309. [CrossRef]
46. Daglar, H.; Erucar, I.; Keskin, S. Recent Advances in Simulating Gas Permeation through MOF Membranes. *Mater. Adv.* **2021**, *2*, 5300–5317. [CrossRef] [PubMed]
47. Kadioglu, O.; Keskin, S. Efficient Separation of Helium from Methane Using MOF Membranes. *Sep. Purif. Technol.* **2018**, *191*, 192–199. [CrossRef]
48. Yang, W.; Liang, H.; Peng, F.; Liu, Z.; Liu, J.; Qiao, Z. Computational Screening of Metal–Organic Framework Membranes for the Separation of 15 Gas Mixtures. *Nanomaterials* **2019**, *9*, 467. [CrossRef]
49. Daglar, H.; Aydin, S.; Keskin, S. MOF-Based MMMs Breaking the Upper Bounds of Polymers for a Large Variety of Gas Separations. *Sep. Purif. Technol.* **2022**, *281*, 119811. [CrossRef]
50. Ongari, D.; Yakutovich, A.V.; Talirz, L.; Smit, B. Building a Consistent and Reproducible Database for Adsorption Evaluation in Covalent–Organic Frameworks. *ACS Cent. Sci.* **2019**, *5*, 1663–1675. [CrossRef]
51. Aydin, S.; Altintas, C.; Keskin, S. High-Throughput Screening of COF Membranes and COF/Polymer MMMs for Helium Separation and Hydrogen Purification. *ACS Appl. Mater. Interfaces* **2022**, *14*, 21738–21749. [CrossRef]
52. Feng, M.; Cheng, M.; Deng, J.; Ji, X.; Zhou, L.; Dang, Y.; Bi, K.; Dai, Z.; Dai, Y. High-Throughput Computational Screening of Covalent−Organic Framework Membranes for Helium Purification. *Results Eng.* **2022**, *15*, 100538. [CrossRef]
53. Sava Gallis, D.F.; Parkes, M.V.; Greathouse, J.A.; Zhang, X.; Nenoff, T.M. Enhanced O_2 Selectivity versus N_2 by Partial Metal Substitution in Cu-BTC. *Chem. Mater.* **2015**, *27*, 2018–2025. [CrossRef]
54. Chowdhury, P.; Bikkina, C.; Meister, D.; Dreisbach, F.; Gumma, S. Comparison of Adsorption Isotherms on Cu-BTC Metal Organic Frameworks Synthesized from Different Routes. *Microporous Mesoporous Mater.* **2009**, *117*, 406–413. [CrossRef]
55. Saha, D.; Bao, Z.; Jia, F.; Deng, S. Adsorption of CO_2, CH_4, N_2O, and N_2 on MOF-5, MOF-177, and Zeolite 5A. *Environ. Sci. Technol.* **2010**, *44*, 1820–1826. [CrossRef]
56. Kloutse, F.A.; Hourri, A.; Natarajan, S.; Benard, P.; Chahine, R. Experimental Benchmark Data of CH_4, CO_2 and N_2 Binary and Ternary Mixtures Adsorption on MOF-5. *Sep. Purif. Technol.* **2018**, *197*, 228–236. [CrossRef]
57. Li, T.; Chen, D.-L.; Sullivan, J.E.; Kozlowski, M.T.; Johnson, J.K.; Rosi, N.L. Systematic Modulation and Enhancement of CO_2:N_2 Selectivity and Water Stability in an Isoreticular Series of Bio-MOF-11 Analogues. *Chem. Sci.* **2013**, *4*, 1746. [CrossRef]
58. Cmarik, G.E.; Kim, M.; Cohen, S.M.; Walton, K.S. Tuning the Adsorption Properties of UiO-66 via Ligand Functionalization. *Langmuir* **2012**, *28*, 15606–15613. [CrossRef]
59. Dubskikh, V.A.; Lysova, A.A.; Samsonenko, D.G.; Lavrov, A.N.; Kovalenko, K.A.; Dybtsev, D.N.; Fedin, V.P. 3D Metal–Organic Frameworks Based on Co(II) and Bithiophendicarboxylate: Synthesis, Crystal Structures, Gas Adsorption, and Magnetic Properties. *Molecules* **2021**, *26*, 1269. [CrossRef]
60. Mayo, S.L.; Olafson, B.D.; Goddard, W.A. DREIDING: A Generic Force Field for Molecular Simulations. *J. Phys. Chem.* **1990**, *94*, 8897–8909. [CrossRef]
61. Bai, P.; Tsapatsis, M.; Siepmann, J.I. TraPPE-Zeo: Transferable Potentials for Phase Equilibria Force Field for All-Silica Zeolites. *J. Phys. Chem. C* **2013**, *117*, 24375–24387. [CrossRef]
62. Vujić, B.; Lyubartsev, A.P. Transferable Force-Field for Modelling of CO_2, N_2, O_2 and Ar in All Silica and Na^+ Exchanged Zeolites. *Model. Simul. Mater. Sci. Eng.* **2016**, *24*, 045002. [CrossRef]
63. Pham, T.D.; Xiong, R.; Sandler, S.I.; Lobo, R.F. Experimental and Computational Studies on the Adsorption of CO_2 and N_2 on Pure Silica Zeolites. *Microporous Mesoporous Mater.* **2014**, *185*, 157–166. [CrossRef]
64. Predescu, L.; Tezel, F.H.; Chopra, S. Adsorption of Nitrogen, Methane, Carbon Monoxide, and Their Binary Mixtures on Aluminophosphate Molecular Sieves. *Adsorption* **1997**, *3*, 7–25. [CrossRef]
65. Li, S.; Chen, J.; Wang, Y.; Li, K.; Li, K.; Guo, W.; Zhang, X.; Liu, J.; Tang, X.; Yang, J.; et al. Adsorption and Separation of CH_4/N_2 by Electrically Neutral Skeleton AlPO Molecular Sieves. *Sep. Purif. Technol.* **2022**, *286*, 120497. [CrossRef]
66. Zhao, Z.; Ma, X.; Li, Z.; Lin, Y.S. Synthesis, Characterization and Gas Transport Properties of MOF-5 Membranes. *J. Membr. Sci.* **2011**, *382*, 82–90. [CrossRef]
67. Cao, F.; Zhang, C.; Xiao, Y.; Huang, H.; Zhang, W.; Liu, D.; Zhong, C.; Yang, Q.; Yang, Z.; Lu, X. Helium Recovery by a Cu-BTC Metal–Organic-Framework Membrane. *Ind. Eng. Chem. Res.* **2012**, *51*, 11274–11278. [CrossRef]

68. Kasik, A.; Dong, X.; Lin, Y.S. Synthesis and Stability of Zeolitic Imidazolate Framework-68 Membranes. *Microporous Mesoporous Mater.* **2015**, *204*, 99–105. [CrossRef]
69. Liu, Y.; Zhang, B.; Liu, D.; Sheng, P.; Lai, Z. Fabrication and Molecular Transport Studies of Highly C-Oriented AFI Membranes. *J. Membr. Sci.* **2017**, *528*, 46–54. [CrossRef]
70. Nishiyama, N.; Ueyama, K.; Matsukata, M. Gas Permeation through Zeolite-Alumina Composite Membranes. *AIChE J.* **1997**, *43*, 2724–2730. [CrossRef]
71. Zhang, P.; Gong, C.; Zhou, T.; Du, P.; Song, J.; Shi, M.; Wang, X.; Gu, X. Helium Extraction from Natural Gas Using DD3R Zeolite Membranes. *Chin. J. Chem. Eng.* **2021**, *49*, 122–129. [CrossRef]
72. Robeson, L.M. Correlation of Separation Factor versus Permeability for Polymeric Membranes. *J. Membr. Sci.* **1991**, *62*, 165–185. [CrossRef]
73. Wu, A.X.; Drayton, J.A.; Smith, Z.P. The Perfluoropolymer Upper Bound. *AIChE J.* **2019**, *65*, e16700. [CrossRef]
74. Willems, T.F.; Rycroft, C.H.; Kazi, M.; Meza, J.C.; Haranczyk, M. Algorithms and Tools for High-Throughput Geometry-Based Analysis of Crystalline Porous Materials. *Microporous Mesoporous Mater.* **2012**, *149*, 134–141. [CrossRef]
75. Dubbeldam, D.; Calero, S.; Ellis, D.E.; Snurr, R.Q. RASPA: Molecular Simulation Software for Adsorption and Diffusion in Flexible Nanoporous Materials. *Mol. Simul.* **2016**, *42*, 81–101. [CrossRef]
76. Dubbeldam, D.; Calero, S.; Vlugt, T.J.H. IRASPA: GPU-Accelerated Visualization Software for Materials Scientists. *Mol. Simul.* **2018**, *44*, 653–676. [CrossRef]
77. Potoff, J.J.; Siepmann, J.I. Vapor–Liquid Equilibria of Mixtures Containing Alkanes, Carbon Dioxide, and Nitrogen. *AIChE J.* **2001**, *47*, 1676–1682. [CrossRef]
78. Jhirschfelder, J.O.; Curtiss, R.B. *Bird Molecular Theory of Gases and Liquids*; Wiley: New York, NY, USA, 1954.
79. Rappe, A.K.; Casewit, C.J.; Colwell, K.S.; Goddard, W.A.; Skiff, W.M. UFF, a Full Periodic Table Force Field for Molecular Mechanics and Molecular Dynamics Simulations. *J. Am. Chem. Soc.* **1992**, *114*, 10024–10035. [CrossRef]
80. Campañá, C.; Mussard, B.; Woo, T.K. Electrostatic Potential Derived Atomic Charges for Periodic Systems Using a Modified Error Functional. *J. Chem. Theory Comput.* **2009**, *5*, 2866–2878. [CrossRef]
81. Manz, T.A.; Limas, N.G. Introducing DDEC6 Atomic Population Analysis: Part 1. Charge Partitioning Theory and Methodology. *RSC Adv.* **2016**, *6*, 47771–47801. [CrossRef]
82. Sladekova, K.; Campbell, C.; Grant, C.; Fletcher, A.J.; Gomes, J.R.B.; Jorge, M. The Effect of Atomic Point Charges on Adsorption Isotherms of CO_2 and Water in Metal Organic Frameworks. *Adsorption* **2019**, *26*, 663–685. [CrossRef]
83. Grenev, I.V.; Klimkin, N.D.; Shamanaeva, I.A.; Shubin, A.A.; Chetyrin, I.A.; Gavrilov, V.Y. A Novel Adsorption-Based Method for Revealing the Si Distribution in SAPO Molecular Sieves: The Case of SAPO-11. *Microporous Mesoporous Mater.* **2021**, *328*, 111503. [CrossRef]
84. Grenev, I.V.; Shubin, A.A.; Solovyeva, M.V.; Gordeeva, L.G. The Impact of Framework Flexibility and Defects on the Water Adsorption in CAU-10-H. *Phys. Chem. Chem. Phys.* **2021**, *23*, 21329–21337. [CrossRef] [PubMed]
85. Nazarian, D.; Camp, J.S.; Sholl, D.S. A Comprehensive Set of High-Quality Point Charges for Simulations of Metal–Organic Frameworks. *Chem. Mater.* **2016**, *28*, 785–793. [CrossRef]
86. Kancharlapalli, S.; Gopalan, A.; Haranczyk, M.; Snurr, R.Q. Fast and Accurate Machine Learning Strategy for Calculating Partial Atomic Charges in Metal–Organic Frameworks. *J. Chem. Theory Comput.* **2021**, *17*, 3052–3064. [CrossRef] [PubMed]
87. Chung, Y.G.; Gómez-Gualdrón, D.A.; Li, P.; Leperi, K.T.; Deria, P.; Zhang, H.; Vermeulen, N.A.; Stoddart, J.F.; You, F.; Hupp, J.T.; et al. In Silico Discovery of Metal-Organic Frameworks for Precombustion CO_2 Capture Using a Genetic Algorithm. *Sci. Adv.* **2016**, *2*, e1600909. [CrossRef] [PubMed]
88. Frenkel, D.; Smit, B. *Understanding Molecular Simulation: From Algorithms to Applications*; Academic Press: Bodmin, UK, 2001; ISBN 978-0-08-051998-2.

MDPI AG
Grosspeteranlage 5
4052 Basel
Switzerland
Tel.: +41 61 683 77 34

Molecules Editorial Office
E-mail: molecules@mdpi.com
www.mdpi.com/journal/molecules

www.ingramcontent.com/pod-product-compliance
Lightning Source LLC
Chambersburg PA
CBHW062059220926
51850CB00010B/2820